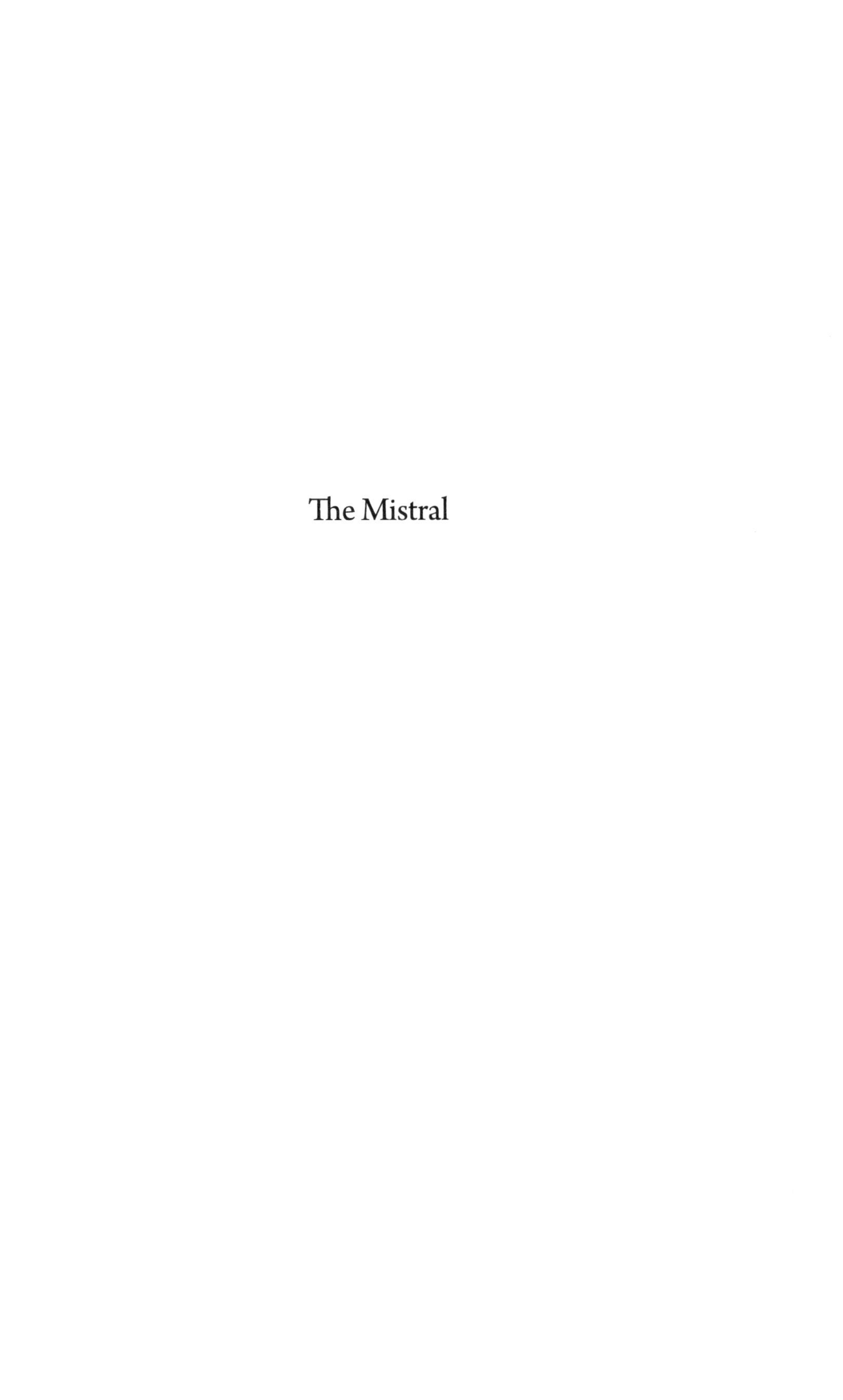

The Mistral

The Mistral

A WINDSWEPT HISTORY OF MODERN FRANCE

Catherine Tatiana Dunlop

The University of Chicago Press   Chicago and London

The University of Chicago Press, Chicago 60637
The University of Chicago Press, Ltd., London

For more information, contact the University of Chicago Press, 1427 E. 60th St., Chicago, IL 60637.
Published 2024
Printed in the United States of America

33 32 31 30 29 28 27 26 25 24   1 2 3 4 5

ISBN-13: 978-0-226-82754-4 (cloth)
ISBN-13: 978-0-226-82755-1 (e-book)
DOI: https://doi.org/10.7208/chicago/9780226827551.001.0001

Frontispiece: François Huard, *Haie de peupliers sous le vent*, 1824. Pen and ink on paper. © S. Normand - Cd13, Museon Arlaten–musée de Provence.

Library of Congress Cataloging-in-Publication Data

Names: Dunlop, Catherine Tatiana, author.
Title: The mistral : a windswept history of modern France / Catherine Tatiana Dunlop.
Description: Chicago : The University of Chicago Press, 2024. | Includes bibliographical references and index.
Identifiers: LCCN 2023054303 | ISBN 9780226827544 (cloth) | ISBN 9780226827551 (ebook)
Subjects: LCSH: Mistral—History—19th century. | Winds—France—Provence—History—19th century. | Human ecology—France—Provence—History—19th century. | Atmospheric science—France—History—19th century.
Classification: LCC QC939.M57 D85 2024 | DDC 551.51/85—dc23/eng/20231222
LC record available at https://lccn.loc.gov/2023054303

♾ This paper meets the requirements of ANSI/NISO Z39.48-1992 (Permanence of Paper).

For Erick Johnson

Windswept *(adj.): (1) Exposed to strong winds*
(2) Made messy by the wind

CONTENTS

Introduction

When the mistral sweeps across Provence, it brings a wild and restless energy to the world-famous landscapes of southern France. Carefully tended farm fields become scenes of total disorder. Snapped vines hang lifelessly off trellises and unripened fruits lie bruised and strewn about the ground. Stalks of wheat that typically stand strong and upright fold down together into a single mass, becoming an inland sea of golden waves. Meanwhile, Provence's actual sea, the ancient and storied Mediterranean, loses its azure color as the mistral pushes against its surface, cloaking its churning waters with a layer of white foam. Near the limestone cliffs that fall into the sea, clouds of dust and pine tree pollen dance frantically about, painting the air with dashes of yellow. The only place to find serenity in this windswept landscape is above you. Even as it unleashes chaos and destruction below, the mistral cleanses and purifies the sky, chasing the clouds away and leaving a brilliant clear blue atmosphere in its wake.

Seizing and jostling everything in its path, the mistral produces an impressive range of sounds. It finds the cracks between roof tiles and around windowpanes, turning habitable structures into musical instruments that whistle and shake under its force. French novelist Émile Zola likened the mistral's presence in a *mas*, a traditional stone farmhouse of the region, to an unseen visitor that "moaned and sobbed wildly" as it made its way through hallways, slamming doors as it slipped through the building.[1] When nineteenth-century botanist Charles Martins found himself immersed in the mistral near the summit of Mont Ventoux, he recalled "a noise like an artillery detonation that seemed to rattle the mountain down to its foundations."[2] Further to the south, at the Old Port of Marseille, writer Alphonse Daudet likened the mistral to a symphony conductor that took all the clamors from Mediterranean ships

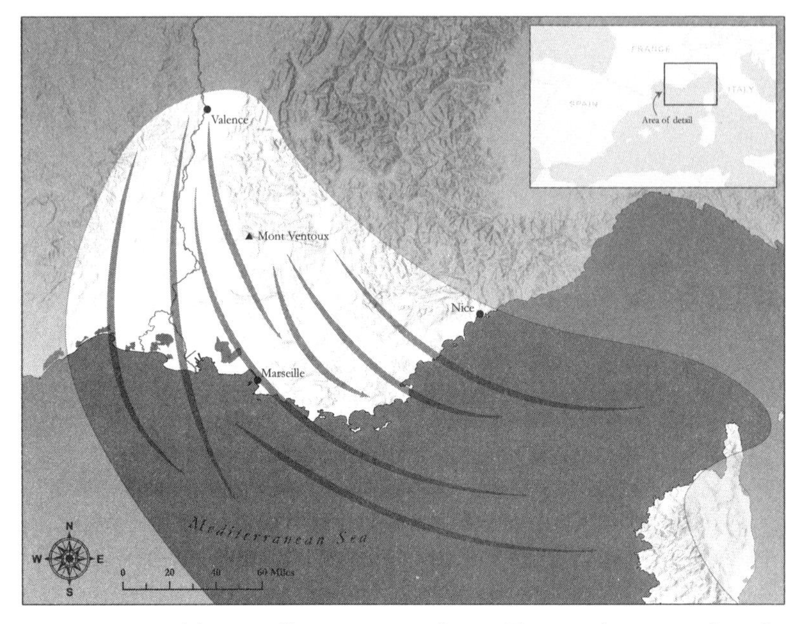

Figure 0.1. Map of the mistral's maximum wind zone. The mistral originates from the inland mountains in south-central France and gusts downward through mountain valleys, open plains, and coastal wetlands, eventually terminating over the Mediterranean Sea. Created by Adam Creitz, Geospatial Core Facility, Montana State University–Bozeman.

and crews and "rolled them up, drove them, mixed them up with its own voice, and made of them a music that was crazy, wild, and heroic."[3]

On a bigger scale, we can think of the entire geography of Provence as a vocal cord system in which the mistral's icy breath descends from the frigid peaks of the Massif Central, pushes through the narrow throat of the Rhône River valley, and bursts forth with a tremendous howl as it opens up into the plains of the Crau, the wetlands of the Camargue, and the open waters of the Mediterranean Sea (see figures 0.1 and 0.2).[4] The part of the sea that surrounds Marseille, the Gulf of Lion, purportedly got its name from the "lion's roar" of the mistral.[5] Some have even likened the cadence of Provençal, the regional language spoken in Provence, to the sound of the mistral itself. "This dialect," declared the Marseillais writer Victor Gelu, "is brutal and impetuous like the wind from the northwest that gave birth to it and imprinted it with its storm-like cachet."[6]

It is not easy for people to withstand the sheer physical force of the mistral. Ex-votos hung on church walls across the South of France thank God for sparing the lives of travelers whose carriages crashed to the ground or whose boats sank to the depths of the sea in the midst of the mistral's roar. In his travel writing, renowned eighteenth-century

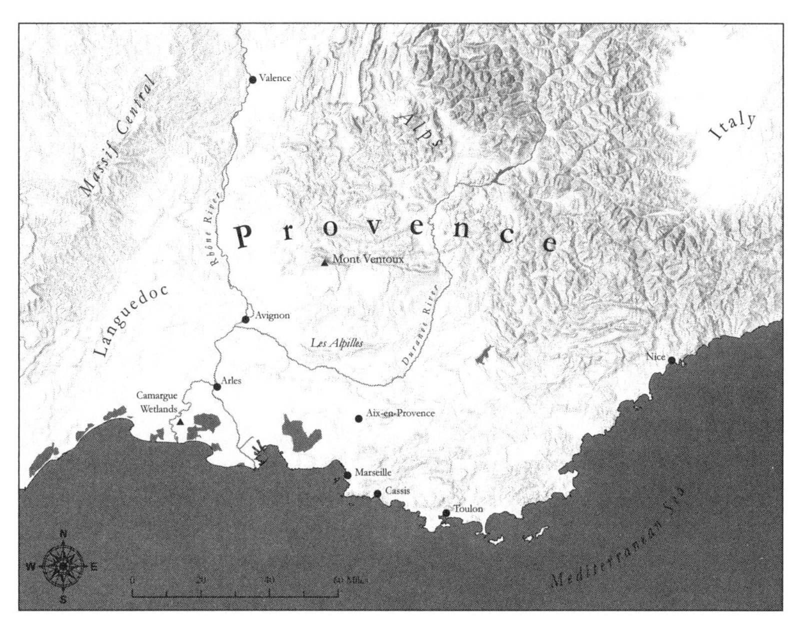

Figure 0.2. Map of Provence. While this historic region lost its official status during the French Revolution, it remained, in the minds of many Provençaux and tourists from the outside, a distinctive part of France with a unique environmental character as well as a special culture. Created by Adam Creitz, Geospatial Core Facility, Montana State University–Bozeman.

mountain climber Horace Bénédict de Saussure—a man who was no stranger to extreme environments—recalled how he was nearly swept away by a gust of the Provençal wind while visiting the ancient city of Arles. The experienced Alpine climber was standing on a rooftop terrace, admiring the city's Roman arena, when "a puff of the mistral with an extreme violence seized me without warning." The wind, he recalled, "would have thrown me into the street" if he had not clung to a chimney on the slope of the roof.[7] Another famous visitor to Arles, the outcast modern artist Vincent van Gogh, called the task of painting outdoors in the mistral nothing less than "the devil's own job."[8] In order to keep his canvases from flying away in the wind, he had to hammer his easels into the limestone soil or pin them to the ground with his hands and knees.

As van Gogh's paintings beautifully demonstrate, the natural world, too, bows down to the invisible force of the mistral. Across the Provençal countryside, pine trees are frozen in a backward-leaning pose, their trunks and branches arching dramatically to one side. The wind may be temporarily gone, but the trees are forever marked by their encounters with the mistral, stuck in a position of futile resistance (see figure 0.3).[9]

Figure 0.3. Coastal pine trees near the Calanques National Park in Cassis. The flag-like structure of the trees results from frequent gusts of the mistral. Photo by author.

Nature's other forms of bending are more ephemeral. When the mistral blows, water in the Bays of Cassis, Toulon, and Marseille moves away from the shore rather than toward it, creating reverse wave patterns in the sea. Not to be forgotten are the animals—the water birds and wild horses who cluster their bodies together in the Camargue wetlands for protection when the mistral is gusting. Popular proverbs—"it takes a mistral to pull the tail off the donkey" and "it takes a mistral to dehorn the bulls"—point to the mistral's power over animal life.[10]

Altogether, the mistral's harrowing impact on bodies and landscapes reveals a side of Provençal nature that is far more ferocious than the region's heavily marketed image as a warm, inviting, and sumptuous vacationland. Blowing for approximately a third of the year, at speeds that often exceed sixty miles per hour, the mistral gives Provence a Janus-faced quality.[11] "Poetic Provence is nonetheless a savage landscape," the nineteenth-century French historian Jules Michelet warned his readers, asserting, "The gusts of winds, brisk and powerful, can have a fatal grasp . . . It is a nature that is capricious, passionate, angry, and charming."[12] Joseph Conrad, in his novel set in Marseille, used the mistral to emphasize the surprising toughness of a place that on its surface appeared soft and temperate. "The mistral howled in the sunshine," Conrad observed, “shaking the bare bushes quite furiously. And everything was bright and hard, the air was hard, the light was hard, the ground under our feet was hard.”[13]

Unleashing the mistral’s hidden power from the dusty corners of the archives, this book is about an unstoppable, unruly force of nature that routinely filled Provence with its mighty presence during the century following the Revolution of 1789 and, in so doing, pushed back against the centralizing power of the French nation-state. The mistral’s violent and restless materiality prevented France’s modern political and economic leadership from establishing the kind of permanent technological infrastructure, efficient national economy, and moderate climate conditions that they desired in Provence. Meanwhile, for those Provençaux seeking greater autonomy from Paris, the mistral’s ability to overpower central planning and disrupt modernization schemes turned it into a model for regionalist resistance. By the end of the nineteenth century, the Provençal population understood the mistral to be something much more than just the weather; it had become a cherished part of their regional identity. Challenging and shaping both nation- and region-building projects in Provence, the mistral emerged as a powerful nonhuman force for historical change during a crucial turning point of modernity.

Placing History in the Mistral Windscape

For all its power to shape, and even dominate, the lives of people in Provence, the mistral is missing from most accounts of French history.[14] What explains its absence? By and large, scholars have tended to view the spaces of history from an anthropocentric perspective, downplaying the unruly weather features of historical landscapes while emphasizing people’s capacity to govern their surroundings. This human-centered approach to the past is often reinforced by the historical documents that scholars use. Most of the old maps found in French archives depict the territory of Provence through the lens of its shifting human-made geographies: as a *provincia* in the Roman Empire, as a trading center in the sixteenth-century Mediterranean world, or as a modern administrative region locked inside the borders of a centralized French nation-state. These anthropocentric maps promote the misleading illusion that natural environments existed merely as backdrops to history—nonthreatening, malleable, and passive—while human beings actively transformed territory according to their will. Together with the entire human-generated historical archive, maps of European territory have helped to silence a reality that is becoming all too clear in the age of climate change: natural environments and weather systems exert their own dynamic pressures on history, whether people like it or not.[15]

By shedding light on the active natural forces that have transformed European territory and the societies that inhabited it, this study seeks to fundamentally reframe our understanding of who (or what) has the power to shape history. This book focuses our attention on a natural geography that has always existed within the confines of a broader French polity, but whose energetic presence never fit neatly into the governing structures of the French state. The mistral windscape presents an intriguing new kind of geographic framework for historians: a natural region that had no fixed boundaries or legal jurisdiction but one that generated a distinctive sensory environment that people felt, experienced, and embodied from the bottom up.[16]

Unlike the trade winds that sweep across vast open oceans, the mistral is a classic example of an inland-generated "local wind," whose direction, temperature, and speed are all determined by the unique shape of regional landforms.[17] Its movement through the southern French landscape is stimulated when an imbalance occurs between a high-pressure zone in the mountainous areas of central France and a low-pressure zone over the Mediterranean Sea. When such a pressure imbalance forms, particularly in the winter and the spring, the cold air sitting atop the Massif Central and the Alps begins to stream downward toward the Gulf of Lion. As it descends, the mistral's air molecules journey through tight mountain valleys, which compress the wind like a garden hose, further enhancing its intensity. By the time the mistral reaches the boundless terrain of the Rhône valley and the Camargue wetlands, its mountain-generated air can feel like a hurricane.

Crucial to understanding the mistral's role in history is recognizing that its windscape is not simply an ephemeral zone of air movement that comes and goes with the seasonal cycles of the Mediterranean climate. To the contrary, the mistral's windscape has become *part of* France's internal physical makeup, concretized, over the centuries, in the biogeography of southern France: in the flora and fauna that have successfully adapted to the wind-impacted space. From the tiny low-lying plants that sprout from the stony peaks of Mont Ventoux to the cliffside bird dwellings of the Alpilles, to the fish that swim in the salt-encrusted deltas of the Camargue, every living thing in the South of France is bound into a web of coexistence with the mistral's northwesterly gusts.[18] Like the animals and plants around them, human beings, too, have rendered the mistral windscape visible over time through their built environment. From their mas farmhouses to their towering cypress hedgerows, Provençaux have actively shaped the wind-adapted look of the region that they inhabit. The mistral windscape thus offers an ideal geographic frame for exploring the day-to-day relationships between a regional community of Europeans and the volatile zone of air that has surrounded them.

A hundred years ago, the Annales school, a movement of historical thought founded in France during the 1920s, attempted—with some successes and some notable failures—to merge the study of geography with the study of history. Influenced by a dark time when the flames of nationalism raged across Europe, Annales school scholars sought refuge and solace in the long-term material and nonhuman factors that shaped history, rather than in the flashy yet fleeting history of its political regimes. In his late-in-life study, *The Identity of France*, the school's leading student, Fernand Braudel, set out to explore "the relations—multiple, intertwined, elusive—between the history of France and the physical territory by which it is confined, sustained and in a way (though not of course completely) explained."[19] Significantly, Braudel understood France not as singular place, but as a collection of places, each shaped by a combination of natural and human factors. His endeavor was to excavate the "many-shaped and multicolored fragments of the mosaic that is France."[20]

Building on Braudel's goal of elevating the material reality of France in the story of its past—while invigorating his approach to history with more recent scholarly perspectives on body, place, climate, and the environment—this book presents a fresh, multilayered view of French geography that sees naturally occurring regional ecosystems and subnational regional communities as deeply entangled and engaged in a constant state of interaction. The result is a study that brings Provence's blustery southern windscape—a subject that has primarily attracted the attention of scientists and climate modelers—into dialogue with France's wider political, economic, scientific, and cultural transformations during the century following the French Revolution. My analysis hinges on multiple overlapping territories: an empowered state space managed from Paris, a politically defunct yet culturally influential southern French region, and a naturally occurring windscape with no official borders. Throughout the book, these three distinct kinds of territory overlap, intersect, and clash, making for a novel analysis of French center-periphery relations that integrates the dynamism of regional environments.

The Origins of France's Master Wind

Compared to the young nation-states taking shape in nineteenth-century Europe, the mistral's territorial presence is incredibly ancient. According to geologists, the mistral first appeared on Earth nearly 3.2 million years ago, during the mid to late Pliocene era, when the upward thrust of tectonic plates produced new mountain ranges that altered the flow of the air in their vicinity.[21] The mistral's genesis corresponded with the creation of what European scientists now term the Mediterranean Biogeographical Region (see plate 1). Extending across a broad swath of the Mediterranean zone, in a manner that cuts through national boundaries, this ecological region is characterized by dry, warm summers and cool, wet winters. Since the Pliocene era, the Mediterranean Biogeographical Region has been home to a raucous family of small-scale winds—the mistral, the tramontane, and the sirocco, among others—whose powerful gusts form a key component of the Mediterranean climate system.

The first human documentation of the mistral did not appear until the sixth century BC, when Greek sailors landed on the northern coast of the Mediterranean and built a settlement that they named Massalia (now Marseille). Calling the wind surrounding their outpost the *melamboreas*, or the black Boreas, the ancient Greek settlers brought with them a mythological understanding of winds as mighty supernatural forces that exerted great power over human beings. Personified as a gruff god with a full beard and flowing robes, Boreas was feared and respected by the Greeks. In his second-century *Geography*, their leading geographer, Strabo, referred to the *melamboreas* as a wind so mighty that it "overturns chariots and strips men of their weapons and their clothes."[22] Later, when the Romans conquered Provence, they replaced the name Boreas with their own term: *circius*. Rather than invoking masculine gruffness, the term *circius* expressed the wind's disorienting, whirling quality.[23] The Roman writer Seneca described the *circius* as a wind that ravaged Gaul and shook its buildings, while the emperor Augustus, during his stay in the region, reportedly built a temple in honor of the fierce northwesterly wind.[24]

The current name for the same wind—*le mistral*—entered the French language during the seventeenth century from Provençal, the regional language spoken in parts of southern France. The Provençal terms for the wind—*maistral*, *maistrau*, and *mistrau*—all derive from the Latin word *magistralis* or *magister*, meaning "the master." Significantly, Provence's neighboring Mediterranean coastal polities all used the same term in their own languages to describe a strong, northwesterly wind: *mestral* (Catalan), *maestral* (Spanish), and *maestrale* (Italian).[25] The name "mistral" thus emerged from a period in time when winds played a central role in a thriving early modern Mediterranean economy that brought Catalans, Spanish, Italians, and Provençaux into a maritime trading network. Together with other prominent Mediterranean winds, the mistral structured and constrained the sail-powered movement of goods and people across their shared sea.

Before the social and economic transformations wrought by an intensive modernization period in nineteenth-century Europe, people in Provence—most of whom worked outdoors—conceived of the mistral as a sort of natural "master" to be approached with humility, fear, and respect. The early modern landscape that took shape in Provence can be read as an index of climate adaptations developed for and by working people. Shepherds, millers, farmers, sailors, and fishers developed region-specific tools—ranging from low-lying stone shelters to windmills to thoughtfully crafted boats—to help them move through, cultivate, and shelter themselves and their animals in their windblown surroundings. Well before the age of modern meteorological stations, poor and illiterate people developed their own savvy forms of weather knowledge based on generational experience, muscle memory, and their careful observations of things like leaves, pine cones, and sails that moved in the wind. Approaching the mistral with a mindset of caution, the nonmodern Provençal population did not seek to triumph over their region's unique "weather world,"[26] but rather to use their place-based knowledge to sensibly adapt to it.

New attitudes toward regional environments in nineteenth-century France erased many of the traditional relationships that working people in Provence once shared with the mistral. As France entered a forward-thinking era of fast-paced industrial, technological, and political "advancement," scientific and government elites challenged sustainable regional strategies for living and working in tandem with the mistral, proposing, instead, to use a combination of state power and modern technology to quell the wind's ferocious power. This book uncovers how post-Revolutionary French governments of all stripes—monarchies, republics, and empires alike—took aggressive actions against the "disobedient" force of nature that persistently rattled the southern reaches of their realm.[27] Within the halls of the central French government, the dominant attitude toward the mistral became one of hostility and confrontation, with the goal of liberating the southern French landscape and economy from the throes of its ancient regional wind.

Taming the Mistral? National Unification and the Struggle for Environmental Order

The new policies and practices toward the mistral that emerged in French government circles after the Revolution of 1789 reflected broad national priorities for achieving territorial order through environmental

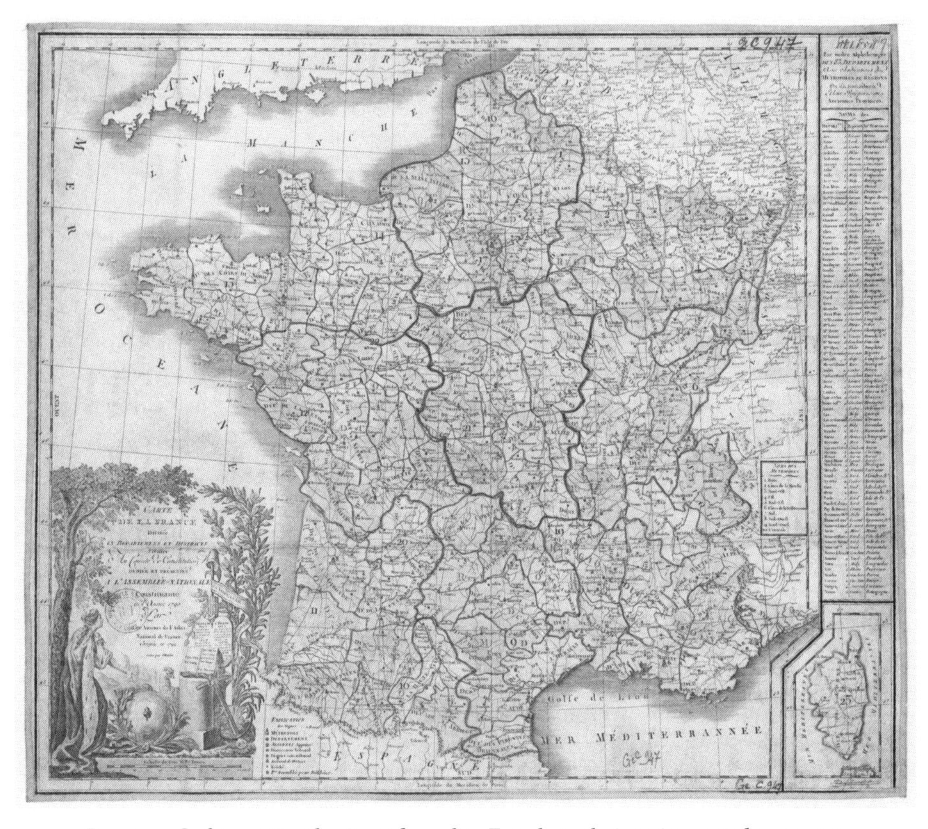

Figure 0.4. Seeking national unity and equality, French revolutionaries created a new map of France in 1790 that replaced historic noble-led provinces with modern administrative units called departments. Over the next century, government officials would use the power of the central state administration to tighten their grip over far-flung provincial populations and their unruly natural environments. *Carte de la France divisée en départements et districts, vérifiée au comité de constitution, dédiée et présentée à l'Assemblée nationale constituante, en l'année 1790, par les auteurs de l'Atlas national de France, corrigée en 1792* (Paris: Dumez, 1792). Bibliothèque nationale de France, département des cartes et plans, GE C-947. Source: gallica.bnf.fr/BnF.

control. Less than a year after revolutionaries stormed the Bastille and upended the social order of Europe's most powerful monarchy, zealous mapmakers redrew the internal political boundaries of France, replacing centuries-old provinces like Provence with new administrative units, called departments, to be managed from Paris (see figure 0.4). By erasing France's noble-led historic provinces from the map of France, the revolutionaries hoped to cement a seamless political geography to mirror a newly unified national community held together by collective principles of liberty, equality, and fraternity.[28] From that point forward, nation-building ambitions in France became closely wedded to the strategic design and alteration of physical territory.[29] But even as post-Revolutionary French governments took meaningful steps toward achieving internal colonization, the country's ecological and climatic diversity remained stubbornly vibrant, dynamic, and real. Like the patchwork of patois-speaking peoples that they encountered across France, post-Revolutionary administrators faced a mosaic of ecoregions that were difficult for central planners to comprehend, much less control.[30]

For the post-Revolutionary French state, the mistral represented instability and disorder. Like the migrating sand dunes of the Landes[31] or the rising floodwaters of the Seine[32]—the mistral came to be viewed as a disruptive force of nature that threatened to impede the nation's economic growth, prosperity, and safety. Whether conservative or republican, nineteenth-century French regimes asserted bureaucratic and technological control over provincial territories in the name of "progress."[33] Beginning under the reign of Napoleon Bonaparte, the central French government used its department administrators to gather and compile environmental knowledge about French regions, dedicating pages of detailed reports to ecological features and weather patterns from all across the country. The topographical indexes and statistical studies that they produced were then placed in the hands of Parisian bureaucracies that sought a legible view of the territories under their management.[34] State ministries and agencies ranging from agriculture to meteorology, maritime navigation to public health, all utilized modern forms of environmental and climatic knowledge to serve the public good through top-down initiatives.

This book excavates today's Provençal landscape as a living physical archive to document the territorial modifications, techno-fixes, and research enterprises undertaken by an activist nineteenth-century French state to make the South of France safer from the mistral.[35] Lining the countryside like sentries are thick, imposing rows of scientifically planned cypress hedges, many of them planted with assistance from experts in the Ministry of Agriculture who advocated for the use of natural barriers to create wind-free microclimates for growing fruits and vegetables for export by rail to national markets. Coastal hillsides around Provence are still covered with thousands of pine trees—the legacy of Emperor Napoleon III's massive state-sponsored afforestation plan to slow the wind's movement through landscape modification.[36] High atop the summit of Mont Ventoux, a hulking concrete weather observatory stands as a reminder of the efforts made by Third Republic meteorologists to replace local climate knowledge with a sophisticated national system of mechanized weather observation.

A burgeoning modern French economy reinforced the state's administrative integration of national territory in the nineteenth century. During this period, the French nation-building project emerged in parallel with the economic consolidation of territorial space through a robust capitalist infrastructure subsidized by political leaders who stood to profit from the accumulation of national wealth.[37] In the metropole as well as across the French overseas empire, steam-powered railway networks, canals, and ports descended like a spider's web, replacing older regionally based forms of transportation, food production, and energy. Traditional wind-centered industries in Provence such as grain milling and small-craft maritime commerce could not withstand the pressures from this emerging large-scale economic system. The crumbling stone windmills scattered across the coastlines and hilltops of late nineteenth-century Provence—their sails torn and their wings broken—became historical vestiges of a place- and community-based economic system that had once flourished in the region and was now obsolete (see figure 0.5).

But even as Provence became enmeshed in national-scale political and economic orders, the mistral's mercurial behavior brutally exposed the limits of French internal colonization. Against all human efforts to render it civilized, predictable, and manageable, the mistral—driven by nothing more than the laws of physics—pushed back with overwhelming violence. In mathematical terms, the force that moving air exerts onto a surface is defined as $F = pA$, with force (F) equaling pressure (p) times the projected surface area facing the wind (A).[38] But nineteenth-century French officials did not have to be experts in physics to grasp the mistral's unstoppable power over Provence. The mistral wrecked every single anemometer that the French National Weather Service installed to track the wind's duration and speed on Mont Ventoux. Nineteenth-century agricultural reports documented dozens of fruit and grain harvests that were devastated by the mistral despite scientific advances in "weather-proof" farming. Steam-powered ships—the ultimate symbols of France's modern industrial economy—met their match when the mistral smashed them like toys against the cliffs of the Mediterranean shore. At a time when the French nation-state was flexing its administrative and economic power over its territory, the mistral became a forceful and bone-chilling reminder that provincial environments were not so easily quelled.

The mistral's natural resistance to human meddling had an intriguing effect on the formation of nineteenth-century regional identity in Provence, a key focus of this book. For those Provençaux wary of the nation-state's increasing power over their *petite patrie*—their "little

Figure 0.5. Postcard of Daudet's windmill in Fontvieille. During the nineteenth century, regionalist writers and artists from Provence elevated windmills in the public imagination from utilitarian structures into Romantic icons of a bygone rural age. The windmill's broken sails were restored by a historic foundation in the twentieth century. © Cd13, Museon Arlaten–musée de Provence.

homeland"—the mistral's ability to destroy mechanisms of outside control turned it into a natural symbol of bottom-up regional defiance. Not unlike the French maquis who turned to their rural scrubland ecosystem as inspiration for their wartime resistance, Provençal regionalists seized upon the mistral's uncontainable power as a metaphor for their own independence and pride. Though Provence's rivers had been canalized and its forests scientifically administered, the air—which the renowned nineteenth-century French geographer Élisée Reclus called "the great atmospheric sea"[39]—evaded the managerial hand of the centralizing French state. Gendered male, *le* mistral—which became known in regional literature as the "king" or the "emperor" of Provence—appeared to safeguard the masculine energy of the Provençal people and protect them from pacification and conquest from the North. The mistral's encounter with the modern French state is therefore not a straightforward story of a national government and its aligned capitalist economy overpowering the wind's natural authority through landscape modifications, techno-fixes, and a network of weather observatories. It is also a story of a growing bottom-up movement, from within the Provençal population, to embrace the mistral's wild power as a beloved form of natural patrimony.

Embracing the Mistral: Nature and the Invention of Regional Identity in Modern France

A defunct territory from the age of monarchy, Provence disappeared from the map of France during the Revolution of 1789. Yet even after it officially ceased to exist, Provence continued to attract the affection and loyalty of many people who kept the ghost of the old province alive through civil associations that preserved its distinctive language, dress, food, traditions, and historical monuments. While scholars have examined the rise of French regionalism through the lens of cultural-preservation efforts, the role of environmental discourse in the formation of regional identities has gone largely unexplored.[40] This book centers nature in the rise of French regionalism, arguing that environmental claims were just as critical to region-building projects as they were to larger French nation-building efforts.[41] Thanks to the bottom-up initiatives of engaged regional citizens, the mistral—like bouillabaisse and bullfighting—transformed into an identity-giving feature of Provence and a wellspring of regional pride (see figure 0.6).

By uncovering the affective bonds that formed between the mistral and regional communities in nineteenth-century Provence, this book invites us to reconsider some of our basic assumptions about the development of territorial identity in post-Revolutionary France. Political

Figure 0.6. A late nineteenth-century postcard frames the mistral's disruptive gusts as an authentic Provençal experience. Archives de Marseille.

scientists have referred to France as a "contractual" nation, in which citizenship is defined by one's willing participation in a democratic system of government.[42] Nineteenth-century historian Ernest Renan called this concept the *plébiscite de tous les jours*: the silent patriotic pledge that French people make each and every day to be loyal citizens of a universal republic headquartered in Paris.[43] Born of human aspirations, this dominant idea of French national identity—which was fundamentally anthropocentric—caused problems when French citizens considered what their "homeland" meant in actual physical terms. "France" was not just an ideal political community or a hexagon-shaped space with adjacent cutouts of its empire pictured on a classroom wall map. In the lived experience of its nineteenth-century citizens, France was also a living, moving, and palpable three-dimensional place: a varied physical territory inhabited by howling ancient winds and centuries-old layers of earth, salt, sand, and water.

In focusing on a *place* called France, and especially its southern reaches in Provence, this book argues that territorial identity, as it came to be understood by ordinary citizens in the nineteenth century, was less contractual, less cerebral, more physical, and more rooted in local and regional scales of living than previously thought. At first glance (or gust), being knocked over by a violent blast of the mistral on a regular basis seems separate from questions of individual or collective identity. It does not involve free will. It does not reinforce the idea of an "imagined national community" of patriots, to use Benedict Anderson's classic framework.[44] And yet, the Provençal population's daily physical encounters with the mistral made them hyperaware of their immediate surroundings, their *environs*. When it swept down from the Massif Central, into the rocky plains of the Crau, and out to the Mediterranean Sea, the mistral forged a regional atmospheric commons that carried the same air from one part of Provence to another, facilitating an embodied regional experience of place for everyone, regardless of their station in life.

Regional environments, in other words, had the power to enter people's bodies and shape their daily experience of life in ways that the *grande patrie*—or the "big country" of France—never could.[45] Sensory experiences of regional weather, I argue, were essential to the development of French citizens' environmental knowledge as well as their environmentally rooted conceptions of home. The mistral's powerful gusts reinforced the fact that Provence did not fit neatly within a national whole; the region's visceral geography, absent on the airless paper map of the nation-state, was real to people through bodily experience.[46] An unofficial territorial consciousness arose in nineteenth-century France that privileged regions—particularly those held together by distinctive "weather worlds" like the mistral—over abstract political visions of a unified national space.[47]

It was not just avid Provençal regionalists who embraced this alternative vision of modern France rooted in regional difference. The spread of railway networks, the development of leisure tourism, and the growing popularity of plein air landscape painting all helped to transform nineteenth-century Provence into a popular travel destination for visitors who wanted to experience a unique part of France.[48] While most came for the sun, the food, and the sea, many visitors ended up face-to-face with the mistral, a far less tranquil but nonetheless exhilarating and authentic part of the Provençal regional experience. In the paintings, literature, and travel writing that they left behind, many of these outsiders documented their striking bodily encounters with southern France's masterly wind. While they did not embrace the same homegrown regionalist agenda as Provençal heritage organizations like the Félibrige, these traveling artists, writers, and tourists contributed to a growing popular enthusiasm for exploring and consuming France's distinctive regional environments.

Chapter Organization

For all the noise and bluster that it generates, the mistral, like many environmental actors, left no voice of its own behind in the historical record. There is no box labeled "mistral" anywhere in France's vast archival system. Everything I learned about the wind's entanglements with nineteenth-century French history was filtered through the variety of things that people from different trades and occupations left behind: their maps, charts, journals, medical reports, scientific observations, tools, paintings, poems, short stories, dwellings, urban plans, and landscape architecture. The visual images printed in this book showcase the multifaceted mistral archive that I built from scratch as I followed the trail of the wind wherever I could find it.

This book is therefore not a "biography" of the mistral, nor does it seek to put forward a scientifically precise climate record of the wind's speeds or durations over a period of time. Rather, this book chronicles the social, political, economic, and environmental developments that transformed French Provence over the course of the nineteenth century through the lens of its powerful, pervasive, and distinctive wind. It explores Europeans' changing relationships with nature in the modern era by focusing on their efforts to manage, alter, and preserve the places closest to them—their regional environments. In so doing, it exposes state efforts to combat the mistral and liberate Provençal society from its influence, but it also shows how local and nonlocal figures embraced the wind's power and identity as an expression of regional authenticity and embodied attachment to place. Throughout the book, I emphasize how the mistral constrained and informed, but never determined, the range of choices available to institutions and individuals in modernizing Provence.[49]

The book begins by examining how the post-Revolutionary French state and its aligned capitalist economy challenged traditional wind-centered lifeways in Provence, both on land and at sea. Before the nineteenth century, most people in Provence labored outdoors, where they developed thoughtful forms of vernacular knowledge for working, moving, and dwelling in the mistral wind zone. In chapters 1 and 2, I delve into the close working relationships between people and wind in premodern Provence—focusing on farmers, shepherds, millers, sailors, and fishers, among others—before turning to the devastating impact that steam power, agricultural reforms, and expanding national economic markets brought to these traditional regional livelihoods. I demonstrate how the collapse of wind-centered industries contributed to an environmentally inflected sense of nostalgia among leading Provençal regionalists, who fixated on antiquated Mediterranean wind roses, crumbling windmills, and stories of windblown noble peasants to protect their provincial identity amid the pressures of nation-state consolidation.

Nothing exemplified the French nation-state's top-down approach to managing the mistral better than its construction of meteorological observatories designed to give Parisian administrators predictive knowledge about the mistral's behavior. In chapter 3, I explore the state's ambitious scheme to build a high-mountain weather observatory on Mont Ventoux, Provence's highest peak, in the late nineteenth century. The information gathered at this station fed directly into the bureaucratic channels of the French National Weather Service, which collated the data to create scientific weather maps offering synoptic views of the forces of nature moving across the national realm. But despite the Parisian bureau's successful dissemination of daily national-scale weather maps, it struggled to accurately represent the direction, speed, and frequency of the mistral's elusive gusts. At Mont Ventoux, the mistral routinely demolished anemometers and battered weather observers' bodies, offering a visceral reminder that provincial environments could challenge the control of the French capital.

The stubborn unruliness of provincial environments, I argue, bolstered an embodied form of subnational territorial identity in modern France that ran counter to the powerful idea of a seamless and unified French nation-state. Chapter 4 demonstrates how place- and climate-based research on bodily health elevated provincial nature in the French public imagination. During the nineteenth century, regional environments became the building blocks for a modern French public health system guided by the localized influence of airs and waters. Historical archives reveal that nineteenth-century French doctors and public health officials were preoccupied with the climatic features of local and regional milieux. For medical experts in Provence, the mistral became a topic of fascination and ultimately an object of praise for its perceived cleansing effects on miasmatic disease particles and pollutants hovering in the southern atmosphere. Theirs was a territorial discourse that anchored French bodies in small-scale weather worlds rather than in a homogeneous national space.

French doctors' interest in the intimate relationship between human bodies and regional environments also spilled over into the visual arts, which is the subject of chapter 5. In this final chapter of the book, I examine how nineteenth-century landscape painters became riveted by the mistral's sensory impacts on their bodies while spending time outdoors in Provence.[50] World-famous modern artists including Claude Monet, Paul Gauguin, and Vincent van Gogh—as well as lesser-known Provençal regional painters such as Émile Loubon—promoted an embodied way of knowing French territory. In contrast to the flattened and tightly controlled images of French geography present on national weather maps, their paintings emphasized the effects of France's lively, diverse, and energizing regional environments in people's everyday lives.

Hidden away in the archives for too long, historical accounts of the mistral and its powerful windscape became disconnected from the larger story of modern France, especially the tensions between national unification and regional particularity. Through a creative approach to historical geography that combines research from official state archives with unofficial archives such as literature, paintings, and architecture, this book revives the fierce blast of the mistral, bringing it back into contact with the past society that it touched. Together, the chapters in this book demonstrate how the mistral became the target of economic, administrative, and technological modernization projects in Provence. In their struggle to tame the mistral's power over human society, however, French government officials faced resistance from the wind itself—which relentlessly bent, overturned, and smashed the things in its path—and from people in Provence whose regional identity became bound up with a force of nature so powerful that it could knock them off their feet.

*

When researching and writing this book in Provence, I received frequent visits from the lead actor in this story. While sifting through historical documents within the protected space of a library or an archive, I could often sense the wind's movements outside. It was unnerving to hear loose shutters banging continuously against exterior walls and window frames whistling and shrieking as a blast of air pushed its way through their unsealed crevices. But it was especially during my days outdoors, on the open road, when I was confronted with the full might of the master of Provence. At the summit of Mont Ventoux, the mistral nearly took my breath away when I tried to walk around the grounds of its high-altitude observatory. Near the seashore, I stood in amazement as the mistral swooped in and transformed the Bay of Cassis from a sailor's paradise into a threatening expanse of whitecaps. While biking around the countryside near Avignon, I had to shelter behind a row of hedges when the mistral pushed against me with such force that it felt like a wall of air. During these moments outdoors, humbled by the material power of the wind, I thought of the aspiring painters in nineteenth-century France who left the safety of their ateliers in search of a direct connection with the landscapes at the center of their studies. My own embodied experience of Provence's lively windscape may not be in the endnotes of this book, but the reader can be assured that the memories of my encounters with the mistral are ever present in this story.

CHAPTER 1

Invisible Sculptor

THE MISTRAL AND THE FORMATION OF THE PROVENÇAL LANDSCAPE

> Like the sea, uniform in spite of its waves, the plain conveys a sense of solitude, of immensity, increased by the mistral, which blows without relaxing and without obstacle and by its powerful breath seems to flatten and so widen the landscape. Everything bends before it. The smallest shrubs keep the imprint of its passage and continue twisted and bent toward the south in an attitude of flight.
>
> ALPHONSE DAUDET, "In Camargue"

On a late-night carriage ride through Provence in 1841, French novelist Alexandre Dumas experienced a terrifying encounter with the mistral. "I had no idea of a land tempest, and indeed did not think such a thing could be," he later recalled in his travelogue, *Pictures of Travel in the South of France*. "I had indeed read in Strabo that the *melamboreus* (such is the name he gives this wind) blew stones about like dust, carried the sheep out of the fields as easily as an eagle would do, and throwing the Roman soldiers from their horses, took away their cloaks and helmets; but I had accounted for these things by the exaggerations of the ancients," Dumas continued. Listening to the mistral hiss as it whipped around his carriage and bent the trees around him "like ears of corn," the humbled traveler wrote that "I was obliged to acknowledge that the master of these countries, for the name it bears is derived from *maestro*, had lost nothing of its power by age."[1]

Dumas was just passing through the Provençal countryside when he experienced the full might of its legendary land tempest. But for inhabitants of Provence whose livelihoods revolved around agriculture, the mistral's presence in their working landscape was a constant challenge. In the Provençal language, it was known as the *manjo fango*, the "mud eater," whose powerful force swept precious water away from fields that were already arid and drought prone. "Season of wind, season of nothing," one popular Provençal proverb proclaimed, while another declared: "The wind is only good for moving ships and mills."[2]

Before the modern era, rural communities in Provence worked strategically alongside the mistral, tailoring their cultivation practices to their windswept climate. Artisans and farmers made creative use of existing topographical features and available natural materials to design unusually shaped protective dwellings—including the mas farmhouse and the Camargue hut—as well as small-scale windbreaks that buffered plots of land from the bending force of the wind. Meanwhile, in areas of Provence with little access to water, stone windmills, perched on hilltops overlooking villages, became an important source of motive power, helping local communities turn wheat into flour and uphold their subsistence food system.

Beginning around the time of the French Revolution, however, traditional wind-centered rural lifeways in Provence began to decline as a national-scale plan to promote scientific agriculture came to fruition. Ambitious Parisian administrations, working in tandem with government officials at the departmental level, embraced a philosophy of environmental management that promised to liberate Provence's agricultural economy from the nuisance of the mistral. Arguing that the mistral was not an unchanging part of nature but rather an environmental problem to be solved, agricultural officials and natural-resource managers invested in massive tree-planting campaigns designed to fix Provence's windy climate through afforestation, so that the region could grow more food, more efficiently. Together with sophisticated new windbreaks—living walls grown from cypresses trees cultivated in modern scientific nurseries—many of the pine forests visible in Provence today stand as historical artifacts of a once-popular nineteenth-century French policy of climate amelioration through landscape design.

The French government's push to modernize Provençal agriculture and to slow the mistral in the name of the public good stirred a powerful emotional reaction among some members of Provençal society. Founded in 1854, the Félibrige—a loosely organized group of upper-middle-class writers and poets—launched a spirited critique of scientific agriculture through nostalgic writings that defended Provence's "authentic" rural landscape, its windy climate, and its traditional lifeways. From the perspective of these regionalists, Provence's windswept atmosphere was to be celebrated as a wellspring of its people's toughness and moral character. In their eyes, the mistral was inseparable from the bodies and the souls of the human beings that it touched; its powerful presence created a sinuous link among sky, community, spirit, language, and soil in Provence that no outside power should take away. Through their creative works, an informal, embodied, and visceral form of modern regional geography was born that offered a countervailing vision of territory to the unitary French nation-state. In unearthing the Félibrige's regionalist environmental thinking, along with concurrent Parisian schemes for top-down environmental control, this chapter reveals the centrality of nature to both nation- and region-building projects in modern France.

The Mistral in the Mediterranean Ecosystem

Before we can address the wind-related adaptations that Provençaux made to their lands over time, we must first understand how the mistral fits into the broader ecosystem of southern France. Part of a greater ecological zone that the European Environmental Agency recently termed the Mediterranean Biogeographical Region, the territory of Provence is composed of a varying band of land adjacent to the sea characterized by hilly landscapes with inland plateaus between low mountain ranges.[3] The region's weather pattern is bimodal, meaning that it is characterized by hot, dry summers and cool, wet winters. Alongside seasonal shifts in temperature and precipitation, the Mediterranean Biogeographical Region is subject to seasonal wind regimes that play a critical role in shaping the type of plant and animal life that is possible in the geographic area. Among the various local winds that cycle into Provence, the mistral leaves the greatest mark, through its frigid and hurricane-like blast of northwesterly mountain air.[4]

When it arrives in Provence during critical periods for plant growth, especially in the spring and fall seasons, the mistral can obliterate plants by snapping and destroying fragile stems, leaves, and buds.[5] In the summer months, the mistral's powerful gusts can compound the deleterious effects of drought and high temperatures by causing water to quickly evaporate. The drying effect of the wind can even cause stomata on leaf surfaces to close, preventing plants from acquiring sufficient moisture to continue normal photosynthesis.[6] In Provence today, there are numerous examples of plants that have developed life-saving adaptations to safeguard against the hazardous effects of exposure to the mistral. High atop Mont Ventoux, a tiny, hardy, low-lying flowering plant called saxifrage—from the Latin for "stone breaker"—appears in a burst of purple color between cracks of limestone rock, where it can withstand the force of sixty-mile-per-hour mistral gusts. Closer to shore, perched atop the stunning gray limestone cliffs of the Calanques, Aleppo pines—with their trunks bent sideways and their branches twisting dramatically into space—are some of the few species of trees that have root systems capable of finding water in such a dry environment.[7]

Like the wild plants that have adapted to these challenging atmospheric conditions, the agricultural societies that have burgeoned in the Mediterranean Biogeographical Region have exhibited a long history of creative adaptations to their arid and windswept surroundings.[8] Ancient water-management techniques such as terracing and irrigation have been practiced for centuries across parts of France, Spain, Italy, North Africa, and the Levant. These human-directed territorial modifications have transformed the terrain to such an extent that ecologists often categorize the Mediterranean zone as a "managed" or a "semi-natural" ecosystem.[9] In Provence, as in most of the regions contained within the Mediterranean Basin, this hybrid ecosystem has favored a triad of mainstay crops—wheat, olives, and grapes—that can be reliably grown in arid conditions with proper techniques. These crops, in turn, have formed the backbone of what is now known as the globally popular "Mediterranean cuisine" that relies on staples such as pasta, olive oil, and wine.

After centuries of careful human sculpting and manipulation, Provence's physical geography had arguably lost much of its "natural" quality by the nineteenth century. But its long history as a hybrid landscape—its mix of natural and human influences—is precisely what makes this region such a fruitful laboratory for examining the shifting practices of French climate adaptation over time.[10] The Mediterranean territory bears the visible imprint of a powerful local wind that has sculpted, eroded, and desiccated its landforms for millennia. But it also exhibits the human markers of a heavily used working landscape that evolved—through bottom-up peasant innovations and top-down government policies—in conversation with its windy environs.[11]

Peasant Life with the Mistral

Discerning which crops were well suited for their regional climate, recognizing cyclical patterns of seasonal change, knowing when to sow seeds, and calculating when to reap the harvest were all long-standing keys to survival for peasant communities in Provence.[12] While most farmers who worked the land before the nineteenth century did not have access to meteorological instruments, they were able to develop their own sophisticated forms of vernacular knowledge about the weather (and its effects on agricultural crops) that they passed down over generations.[13] Unearthing these "bottom-up" forms of peasant environmental knowledge can be a challenge. Unlike government officials and intellectual elites, working people in Provence did not leave behind statistical records or official reports about the weather. Piecing together the real-world working relationships between peasants and the mistral therefore requires us to look for evidence in unofficial types of archives, including oral culture, farmers' journals, and vernacular works of architecture.

In his pioneering investigation into peasant mentalities, historian Robert Darnton underscores the usefulness of proverbs and sayings for shedding light on the forgotten lives of ordinary people. "When we cannot get a proverb, or a joke, or a ritual, or a poem, we know we are on to something," Darnton explains. "By picking at the document where it is most opaque, we may be able to unravel an alien system of meaning. The thread might even lead into a strange and wonderful worldview."[14] In Provence, we are lucky to have an outstanding resource on peasant language in Nobel Prize winner Frédéric Mistral's Provençal-French dictionary, *Lou Tresor dóu Felibrige,* first published in 1871. Like many linguists and ethnographers active in nineteenth-century Europe, Mistral was eager to preserve what he saw as his region's dying traditional peasant culture. In addition to formally translating the Provençal language into French for the first time, his dictionary included numerous proverbs, sayings, idioms, and dicta, many of which reflected a deep relationship between peasant communities and the weather.

Provençal farmers were so physically and mentally intertwined with their regional windscape, Mistral's dictionary reveals, that people believed that they could take on mistral-like characteristics and that the mistral (*lou mistrau* in Provençal) could take on human characteristics. A *coucho-mistrau,* for example, was "a man who exaggerates, or whose words go faster than the wind." Meanwhile, a storm produced by a mistral, a *broufonié du mistrau,* came from the Greek word for a "grave voice," while someone called a *bramareu,* "he who likes to bellow," had the same name as a bellowing wind, or a *mistrau bramareu.* Peasants, moreover, liked to emphasize the wind's visible effects on things that mattered to them. Qualitative descriptions included "The mistral dehorns the oxen"; "The mistral takes the roots off the wheat"; and "the mistral, the poplar-bender." A more specific set of complaints about the mistral—which were less commonly heard in Provence's maritime community but mattered a great deal to farmers—pertained to the issue of the wind's dryness. "The mud eater is aloft, here comes the mistral"; "My hands are dry, a sign of the mistral"; and "The mistral is the direct opposite of rain."[15]

In addition to using metaphors to describe the mistral's sensory qualities, peasants provided detailed explanations of the wind's causes. Several proverbs offered simple rules of thumb for predicting the mistral's arrival: "When the mistral says *Bonjour,* it is here for three, six, or nine days. When it says *Bonsoir*, it is here only until the evening"; "If it begins in daytime, it lasts three days; if it begins in the night, it lasts as long as a bread bakes"; "Fog on the hill presages calm, fog in the valley presages the mistral"; and "The plow that squeaks while working presages the mistral." Other proverbs provided a more mysterious, pagan explanation. "When the cat scratches its ear, the mistral wakes up," referred to the pagan belief that black cats were witches' pets who could change the weather through black magic.[16]

Beyond its frequent appearance in popular proverbs and idioms, regional weather played a central role in shaping the rhythms of traditional farm labor in the Provençal countryside. A daily journal left behind by a nineteenth-century farmer from the district of Arles offers rare insight into the mistral's role in the life of a rural worker. The exact provenance of the journal is unclear, but it appears to have been acquired by the departmental archives of the Bouches-du-Rhône after its owner, a man by the name of Astier, filed for bankruptcy. Reading through his personal entries, it is noticeable that the weather is the very first thing that the farmer writes about each day, before any mention of his work tasks or other activities. A sampling of his daily logs reads:

> March 1st, Saturday
> Very strong mistral has carried everything away and it's cold.
> Worked on the cart and on home improvements.
> March 5th, Wednesday
> Mistral weather and cold.
> Covered at the moment.
> Drove the cart. Plowed behind the
> Mas, laid out the manure with Farmer Martin, Uncle, and Louis.
> Cutting finished.
> June 9th, Monday
> Mistral. Good weather.
> Carting. Finishing the first
> Vines in the old forest
> Began the vine near the canal.
> Day Labor.[17]

The seamless blending of weather information with accounts of daily transactions such as the sale of food at the market underscores the central role of local weather in structuring farmers' sense of time in rural Provence. In addition to their weather-dependent work schedules, rural communities in Provence developed distinctive forms of regional architecture that reflected their close relationships with the mistral.

Signs of Bottom-Up Climate Adaptation in Traditional Provençal Architecture

In contrast to elite architectural forms that travel easily across international borders, such as baroque, neoclassical, or art nouveau styles, vernacular architecture develops out of and tends to remain in distinct geographic spaces.[18] It is this place-based, grounded quality of vernacular architecture that makes it an ideal archive for exploring how working people in Provence once adapted their everyday lives to the mistral. Invented before the modern French state infiltrated its far-flung regions with one-size-fits-all modernization initiatives for built environments, traditional dwellings in Provence reflected what landscape architect Ian McHarg famously called "design with nature": an approach to constructing human habitats that embraces the restrictive conditions of a region's unique weather and topography.[19] Thanks to the region's varied terrain, rural Provençal communities generated a diverse range of vernacular buildings, each carefully adapted to the ecology of their surroundings.

The mas farmhouse, perhaps the most iconic example of vernacular architecture in Provence, was designed specifically for fending off the mistral. Deriving its name from the Latin *mansus*, the stone cottages that lined the Roman roads that once crisscrossed the region, most of the surviving mas farmhouses in Provence today originated in the seventeenth and eighteenth centuries, when agricultural development in Provence expanded from protected hillsides into the region's open plains, meaning that farmers needed to figure out how to build dispersed houses that could withstand exposure to regular wind gusts.[20] The result was distinctive buildings whose odd appearance captured the attention of landscape painters such as Paul Cézanne, a native of Aix-en-Provence, whose paintings such as *Houses in Provence: The Riaux Valley near L'Estaque* beautifully demonstrates a mas's elegant fusion with the geology and vegetation of its natural environment (see plate 2).

Like a person turning their back away from the wind to protect their face, the windward-oriented facade of a mas typically consisted of a windowless fortified wall built low to the ground, while the southern facade of the house had numerous openings: the front entrance, windows, and often a stable door for animals. The two-sided character of the building design, in turn, shaped how it was used.[21] The southward-facing side of a mas, the "sunny side," was where the to-and-fro of everyday life happened in rural Provence, while the northward-facing side was "blind" (*aveugle*), closed off to the world, serving exclusively as a protective rampart against the wind. In addition to employing opposing types of facades, farmers oriented their mas north–south, so that one of the building's corners faced the mistral directly. Called *pointe en avant* (point first), this savvy design strategy could successfully deflect the mistral's invisible force to the left and right by encountering it at an angle.[22] Built from sturdy, locally quarried stones using dry masonry, many centuries-old mas farmhouses have survived until today and continue to weather gusts of wind year after year.

Not all parts of Provence, however, had access to stone building materials to block the mistral. In the Camargue—the coastal region at the mouth of the Rhône River where the mistral reaches some of its fastest speeds—ranchers, fishermen, and salt workers designed wind-proof dwellings from the vegetal materials that grew directly around them in their wetland ecosystem. Reeds, which also form natural defensive barriers from the wind for animals like horses and flamingos, have been used to build homes since the original Greek settlement in the coastal area, over two thousand years ago. "A roof of reeds, walls of reeds, dry and yellow, that is the [Camargue] hut," observed Alphonse Daudet in his nineteenth-century short story "In Camargue."[23]

The reeds (or *sagne*, in Provençal) were thatched together in a simple mode of construction called the *cabane de Camargue* or the *mas de Camargue*. Even inside the home, reed partitions (*cloisons de sagno*) separated the sleeping room from the common room. The regional dwellings exemplify what anthropologist Tim Ingold calls the practice of co-option: when human beings learn from naturally occurring patterns in their environment and design their own structures based on the nature that they see around them.[24] In the late nineteenth century, some of the modern artists who visited the Camargue took the idea of co-opting nature a step further, by integrating vegetal materials from the wetlands into their creative process. During his stay in Provence in 1888–89, Vincent van Gogh harvested his reed pens directly from the wetlands, using them to create pen-and-ink drawings that beautifully evoked the fusion of natural and human habitats in the Camargue (see figure 1.1).

The reed huts scattered across the Camargue wetlands were long, low-to-the-ground buildings that terminated in a distinctive rounded apse turned toward the northwest to face the mistral head on. Like the "point first" directional orientation of the mas farmhouse, the rounded apse of the *cabane de Camargue* was designed to divert air flow away from the main structure and strengthen its stability. It was only in front of the southern-facing facade that fishermen and ranchers would carry out daily tasks such as making nets and mending horse tack. The roof of the hut, meanwhile, was often used to signal the Catholic faith of the

Figure 1.1. In the Camargue wetlands, ranchers and fishers lived in unusually shaped huts made of reeds, or *sagne*. The apses on the north sides of dwellings were rounded to deflect the mistral's gusts. This drawing by Vincent van Gogh illustrates how wetland houses emerged elegantly from a bed of reeds. The drawing itself was made with a reed pen that the artist harvested himself from the site. Ink on paper. Vincent van Gogh, *Landscape with Hut in the Camargue*, 1888. Van Gogh Museum, Amsterdam (Vincent van Gogh Foundation).

family inside. On the roof ridge, builders often affixed a white cross, inclining it slightly backward to let the wind flow over it, as a symbol of divine protection. Sometimes locals would also affix pagan objects such as stuffed owls, bull horns, or shiny stones to their huts in an attempt to ward off bad spirits.[25] Despite these supernatural defenses, however, the feeling of being in a *cabane de Camargue* when the wind was blowing felt, in the words of Alphonse Daudet, like being in the cabin of a boat during a storm at sea:

> Under the assaults of the mistral or the tramontane, the door bursts in, the reeds cry out, and all these little shocks are a mere echo of the great agitations of Nature going on around me. The winter sun lashed by the wind scatters itself, joins its beams, and again disperses . . . light comes in jerks, noises also, and the bells of the flocks heard suddenly, then forgotten, lost in the wind.[26]

Indeed, not unlike a boat, the thatched huts were designed to move and jostle with the wind in order to maintain their structural integrity. Just in case, however, some homeowners tied a central ridge beam to the ground with a cord in order to prevent the hut—with its lightweight reed body—from being blown away.[27]

While working people in Provence typically used Camargue huts and mas farmhouses as their primary dwellings, other forms of vernacular architecture served the needs of laborers on the move. Before their decline over the course of the nineteenth century, seminomadic shepherds used to follow their flocks across the Provençal countryside in a seasonal migration pattern called transhumance.[28] "Shepherds lead a hard and solitary life," explained one expert on rural Provence, "living at all times in the open countryside, exposed to the intemperate air."[29] During the winter months, shepherds would guide their herds of sheep and goats down from the mountainous part of Provence to the warmer plains to feed on the wild shrubs that grow in the cicada-filled garigue landscape, before returning to the mountains when the season changed.[30]

The classic Christmas santon figure "shepherd in the wind" honors the work of this wandering group of pastoral laborers. His body struggling against the mistral, the shepherd stretches his arm above his head to catch his wide-brimmed hat while his *capo*, or cape, made of heavy homespun cloth, flutters behind him (see figure 1.2). To shelter themselves on their seasonal journeys, real-life shepherds turned to rudimentary stone dwellings called *bories*, which resembled stone beehives and dated back to the medieval period. The conical shape of the bories—

Figure 1.2. The Christmas *santon* figure "shepherd in the wind." As fewer and fewer Provençaux earned their livelihoods by working the land, regional elites lamented the disappearing traditional society of the past and its closeness to nature. Photo by author.

lacking any angles—ensured their structural stability in a windy landscape. Comprised of just one ovoid room, a borie could be used as a temporary shelter for a shepherd and his flock to spend the night, or as a storage facility for farmers to secure tools, seeds, and harvested grains. According to one expert, the remains of some four thousand bories have been identified in the Vaucluse, Forcalquier, and Aixois regions of Provence.[31]

While most traditional building types in Provence were designed to block or buffer the mistral, windmills were constructed for the opposite purpose: capturing and harnessing the wind's power. An ancient form of technology that likely arrived in Europe from the Middle East, the first windmills in Provence were built by medieval Templars near Arles in the twelfth century. The buildings spread rapidly across the region in the seventeenth century, especially in areas with rolling, hilly topography and few naturally occurring sources of water.[32] Even when water was available, windmills were sometimes easier to build than watermills because of the complex legal code surrounding water rights. Like sail-powered boats designed to capture the wind and generate motive power, windmills operated on nature's clock. The variability of the atmosphere meant that windmills suffered from a "lack of certainty" and a "fickleness of operation."[33] When threatened by a fierce storm, a miller would often place a cross on the ridgepiece of his entryway to invoke the protection of saints.[34]

Within rural French communities, the mill served as a multiuse "commons" for people to bring their harvested grain to mill, to socialize with each other, and to celebrate religious festivals such as saints' days. "The mill," according to Steven Kaplan, a leading expert on the history of French bread, "was one of the fundamental institutions of old-regime life, along with the marketplace, the church, the court of law, and the tavern."[35] At the centerpiece of the mill's social world were millers (*meuniers*), skilled workers whose livelihoods depended on their firsthand knowledge of the wind. Many millers, in fact, were considered to be the towns' keepers of weather knowledge before meteorology became professionalized in the late nineteenth century.[36] Using a weathervane, millers would attempt to predict the most effective orientation for their sails so that they could best *prendre le vent* (take the wind), not unlike a mariner adjusting his or her boat's sails to capture a favorable gust.

The weather-dependent nature of their jobs meant that millers had little control over their work schedules. According to a professional millers' manual from 1848, Provence's constantly shifting windscape "demands an assiduous miller; a man of sangfroid, whose courage does not escape him in the case of bad weather."[37] A wind miller often worked day and night—twelve to eighteen hours a day or more—seconded by his wife and usually by at least one journeyman or a *garde-moulin*.[38] The miller ate meals in the mill itself, and rarely risked coming down to the village in case the wind changed. Many millers endured physical suffering from the rigors of their labor. The air in the mill itself was unhealthy, containing an "atmosphere thick with dust" that caused ailments such as eye irritation, asthma, and a nasty cough called "miller's wheeze."[39]

Revolutionary changes in economics and technology ushered in by the "progress"-oriented nineteenth century, however, would soon put an end to the wind miller's long days and nights. Wind power's erratic nature made it increasingly incompatible with the needs of France's modern industrial economy, which had seized on steam as a great liberator from the constraints of weather and geography. The construction of Provence's first steam-powered mills, in 1819, marked the beginning of the end not only for the region's wind-powered flour industry but also for the small-scale economic and social order that older mills had held together. As Provence's wind-adapted vernacular buildings—its mills, bories, mases, and huts—slowly collapsed under the southern sun, a massive state-coordinated effort to "fix" the wind and relieve modern France of the burdens of its oppressive regional climates took shape.

Landscapes of Progress: Taming the Mistral?

While traditional Provençal society worked in tandem with the particularities of the southern French climate, the modern French nation-state embraced top-down, dirigiste policies that challenged nature's authority. Rather than accept the gusts of the mistral as normal, a new generation of French government planners asked: What if people could design a landscape capable of slowing, and even eliminating, the mistral? The notion that climate was malleable—that people could change the speed and direction of the wind through landscape "improvements"—became increasingly widespread around the time of the French Revolution. Eager to transform arid territories like Provence into more productive agricultural areas, the French state began to coordinate the "rehabilitation" of the national climate through a massive afforestation scheme that would ultimately result in the annual planting of over twenty-two million new trees across France by the 1860s.[40] Like walls surrounding a castle, the human-generated forests were to serve as robust defenses for French civilization against the mistral. Blurring the lines between nature and artifice, the freshly planted forests raised intriguing questions about a national government's right to alter weather patterns in the perceived interest of the public good.

At the root of the French state's ambitious tree-planting initiative was a changing set of assumptions about the relationship between humans and nature. For centuries, Christian theology had argued that control over the weather lay in the hands of God. The ex-votos found on the walls of French Catholic churches allude to this older mindset toward nature, in which people relied on the intercession of the Virgin Mary and the saints to protect them from hazardous winds and weather. Works of vernacular architecture like windmills and Camargue huts reveal how rural Catholics used the sign of the cross to repel winds. Enlightenment-era scientists and philosophers, on the other hand, rejected the Christian narrative of human helplessness and vulnerability in the face of nature. Through rational thought and empirical observation, "Nature's secrets could be unlocked and adapted to man's purposes."[41]

The mistral was one such "secret" that French philosophers, scientists, and state officials were eager to demystify and bring under human control. In the eyes of Enlightenment-era reformers, the mistral's harmful and water-stealing air was a challenge to the nation's food security and needed to be controlled. For many of these progress-minded thinkers, the mistral's destructiveness was part of a widespread belief in environmental waste and decline that historical geographers have termed "desiccation" or "ruined landscape" theory."[42] Proponents of this theory believed that the Mediterranean climate had worsened over the course of the eighteenth century thanks to the reckless and greedy destruction of the forests that had once blanketed the region. When people deprived the Mediterranean Basin of its tree cover, they inadvertently ushered in a disastrous new era of human-generated climate change in which fast-moving, out-of-control winds started to blow, destroying crops and ruining livelihoods.

In France, one of the strongest advocates of desiccation theory was François-Antoine Rauch, a philosopher whose 1802 treatise on the harmony of nature argued that people, not the Earth's atmosphere or an all-powerful God, had the power to create winds.[43] "The empire of cold or drying winds has grown in every sense," he wrote, "and has changed the general order of vegetation."[44] Like many believers in desiccation theory, however, Rauch was not entirely dismal in his view of human-generated climate change. While he blamed humans for creating the mistral, he also believed in their capacity to fix it. "In reforesting the mountains of the Corbières," he declared optimistically, "we can extinguish the mistral wind . . . In gradually extinguishing the domination of winds that torment us . . . it will be possible, as a consequence, to regenerate and increase the warmth of our climates . . . to bring about seasons that are fixed, regular, and ordered."[45] The first step toward France's climate amelioration, therefore, was figuring out how to bring thousands of new trees to life.

Only the French state—with its substantial financial resources and vast bureaucratic reach—was capable of coordinating a landscape modification project on a such a massive scale. A few years after the publication of Rauch's book, during the reign of Napoleon Bonaparte, climate research became institutionalized within the French government. In contrast to the immediacy of weather, the concept of climate—which encompassed multiyear patterns in the speed and direction of wind, rainfall, and temperature—was appealing to a modernizing French state interested in long-term agricultural planning and prosperity.[46] In Provence, regional climate-improvement initiatives began in the late 1810s, with the distribution of a state-sponsored survey that asked mayors to provide their observations on changes to their local climate.[47] Unlike the "bottom-up" environmental knowledge generated by ordinary shepherds, farmers, ranchers, and fishers, this survey was intended to produce legible statistical information—elite knowledge—capable of guiding the nation-state's top-down, dirigiste landscape planning.

One of the government survey's questions specifically addressed how regional wind patterns had changed over time: "Have the winds been more violent, more destructive, more variable? Have you noticed that the ones from the north or the south appear all at once and with a sudden change of a greater destructiveness than the last century, when France was more wooded?"[48] Local officials' responses to the question were varied. The mayor of Marseille did not notice any particular change in the mistral, noting that it "was already very violent since the time of Augustus."[49] Dr. Gibelin, secretary of the Academy of Aix-en-Provence, on the other hand, claimed that climate change was indeed at work in the steady disappearance of olive trees from his region. "Deforestations, destructions, and cuttings have taken away the natural shelters that our olive trees, our vines, and all of the vegetation"[50] had benefited from in the past.

Echoing this observation, the mayor of Arles wrote that "it is in the cultivation of olive trees that we can note the deadly influence made by the lack of shelters from the meteorological system of this region. This crop has declined by half in the last thirty years."[51] The mayor of Aubagne, a small pottery-making town close to Marseille, humbly admitted that he could not provide a clear answer, explaining that he had "trouble in this region finding people who were educated enough in meteorological science who could communicate their knowledge to me." The best that he could do was to ask old people about changes in the weather. "The elders of this district [*commune*] share the opinion that the deforestation of our mountains and the clearings of our forests are the causes of the floods experienced and the violence of the storms."[52] Despite the patchwork nature of the information that the prefect of the Bouches-du-Rhône in Marseille was able to attain, his conclusion was that deforestation had set Provence's climate on a dangerous course.[53]

Now that he had proof of the problem, the prefect in Marseille, supported by the minister of the interior in Paris, pushed for an aggressive ecological-restoration plan. In search of the perfect tree to plant in Provence, he wrote to the Jardins des Plantes in Paris and Marseille in 1821. These state-sponsored royal botanical gardens had become laboratories for discovering the best tree species to promote the *mise en valeur* of France's uncultivated, or poorly cultivated, lands.[54] After much correspondence, the prefect's office in Marseille narrowed its choice to the pine tree, a species capable of thriving in areas with arid soil. One particularly promising subspecies was the Corsican pine (*Pinus nigra subsp. laricio*), a tree that originated in the high mountains of Corsica and grew abundantly on sandy plains and in rocky areas. Another attractive option was the Aleppo, also called the Jerusalem, pine (*Pinus halepensis*), which was native to both Provence and Algeria and was proven to be "content with the worst terrains."[55]

Over the course of the early to mid nineteenth century, the French national forest administration facilitated the planting of both pine subspecies in Provence with varying degrees of success. Receiving one of the first deliveries of Corsican pine seeds in 1821, the mayor of Cassis wrote that "we transported the case [of seeds] to the summit of Mont-Canaille [today Cap Canaille], and we pegged the seedlings in different locations. Then we watered them. The subsequent dryness, and the winds of the northwest that followed, killed the seedlings. But others survived."[56] By the 1840s, the process of acquiring seeds for reforestation initiatives became increasingly industrialized. The agricultural section of the Royal Academy of Marseille built a *sécherie* (dryer) in Aubagne that employed fire to force pine tree cones to open prematurely and release their seeds.[57] Officials in Aubagne later claimed that locals from other towns had sneaked into their tree nursery and stolen cones to plant on their own hillsides, as a form of unsanctioned bottom-up wind-shelter building.[58]

Alongside the massive young forests of pine trees sprouting up, thousands of new scientifically planned cypress windbreaks were changing the appearance of the nineteenth-century Provençal countryside. "Our region," the poet Frédéric Mistral once reminisced, "has no ramparts other than the large cypress hedges that God made especially for her."[59] Of course, cypress trees did not appear in Provence in a divine act of creation; they were agricultural tools that first arrived in Provence in classical times. The scale of their numbers increased dramatically in the nineteenth century, however, when access to both irrigation technology and railway networks encouraged farmers to switch from a narrow menu of traditional Mediterranean crops—wheat, olives, and wine—to a wide variety of fruits and vegetables that needed a lot of protection from the wind. Between 1840 and 1920, horticulturalists in the lower Rhône valley set out hundreds of scientifically planned rows of cypresses to create protected growing environments whose microclimates were hospitable to plants with delicate stems and leaves.[60] According to one study from 1851, approximately twenty thousand cypresses were planted per year in the Bouches-du-Rhône department.[61]

Like the forestry experts in search of the perfect pine tree, experts in horticulture tested different types of cypresses before settling on the Mediterranean cypress (*Cupressus sempervirens var. pyramidalis*) as the species that was best suited to Provence. In addition to its 80 percent impermeability to the wind, the species was well adapted to poor, stony ground, and it did not invade neighboring crops.[62] Indeed, study after scientific study hailed the usefulness of windbreaks for agricultural productivity. "In the Rhône valley, where the mistral frequently blows, a simple hedge of two meters in height can preserve a distance of twenty-two meters. It is through these types of shelters that we can cultivate peas, melons, and artichokes that do not resist the violence of the winds when they are not protected," explained scientist Antoine-César Becquerel in his study on soil and climate.[63] Echoing this observation, George Perkins Marsh, a leading nineteenth-century American geographer, observed that the shelter belts taking shape in Provence's stark Crau Plain allowed fruits and vegetables to thrive in an area that "had remained a naked waste from the earliest stages of history."[64]

Unlike the expansive forest projects aimed at regularizing the Provençal climate, windbreak construction was a small enough task to be undertaken by individual members of the public. Thanks to rising literacy rates and the advent of cheap printing technology, nineteenth-century Provençal farmers were able to learn about the latest techniques for building "living walls" through the circulation of mass-produced popular manuals and journals written by agricultural experts. One farming manual, for example, suggested that Provençal farmers cultivate cypress-tree nurseries directly on their properties, enabling them to create windbreaks quickly when they were needed.[65] The *Journal d'agriculture pratique*, founded in 1837, frequently offered farmers words of advice about the mistral. Answering a plea from a farmer in the Camargue, one advice column read: "Your property, you say, is blown by the mistral and often the earth is carried away down to 0.10 m in depth. How do you prevent this disaster? Plant a double line of reeds, or *arundo*, as a windbreak . . . The most violent mistral shakes, but does not destroy, it."[66] Another expert wrote: "To prevent the sea salt from dissolving in the water . . . we used to cover the ground with a thick layer of reeds (*Arundo phragmites, L.*) and brown flatsedge (*Cyperus fuscus, L.*), thereby diminishing evaporation during the day from the north winds (mistral)."[67] In his manual on cultivating industrial flowers in Provence, Paul Granger, former head botanist for the French navy, refers to his special "acclimatization garden," which he established in Toulon in 1898 to test mistral-resistant living hedges and screens from as far away as Japan and Portugal.[68]

Thanks to the spread of modern scientific horticulture and the state-sponsored push for national-scale reforestation, a new kind of rural environment took shape in nineteenth-century France, one that still appeared natural—with pine trees and cypresses—but one that was more closely managed by human beings than ever before. To visualize the full extent of this transformation, we need look no further than the changing cover images for the *Journal d'agriculture pratique*. The first image, which appeared on the journal's cover during the 1840s, pictures an arid France that is still dependent on wind power (see figure 1.3). In the foreground, a farmer is struggling to work the land with a rudimentary

MAISON RUSTIQUE DU XIXᴱ SIECLE

Seconde partie.

JOURNAL

D'AGRICULTURE PRATIQUE

ET DE JARDINAGE

PUBLIÉ

PAR LES RÉDACTEURS DE LA MAISON RUSTIQUE DU XIXᵉ SIÈCLE

SOUS LA DIRECTION DE

M. ALEXANDRE BIXIO

Docteur en médecine, membre correspondant de l'Institut national des États-Unis,
de l'Académie royale d'Agriculture de Stockholm,
des Sociétés d'Agriculture de Besançon, Chiavari, Dijon, Lyon, Moulins, etc.

DEUXIÈME SÉRIE — TOME IV

OCTOBRE 1846 A DÉCEMBRE 1847

PARIS

LIBRAIRIE AGRICOLE DE DUSACQ

Éditeur de la *Maison rustique du XIXᵉ siècle* et du *Bon Jardinier*,

RUE JACOB, Nº 26.

1846

Figure 1.3. Cover image from the *Journal d'agriculture pratique et de jardinage* in 1846. The image features an arid landscape near a small village with windmills perched on a hilltop in the background. Source: gallica.bnf.fr/BnF.

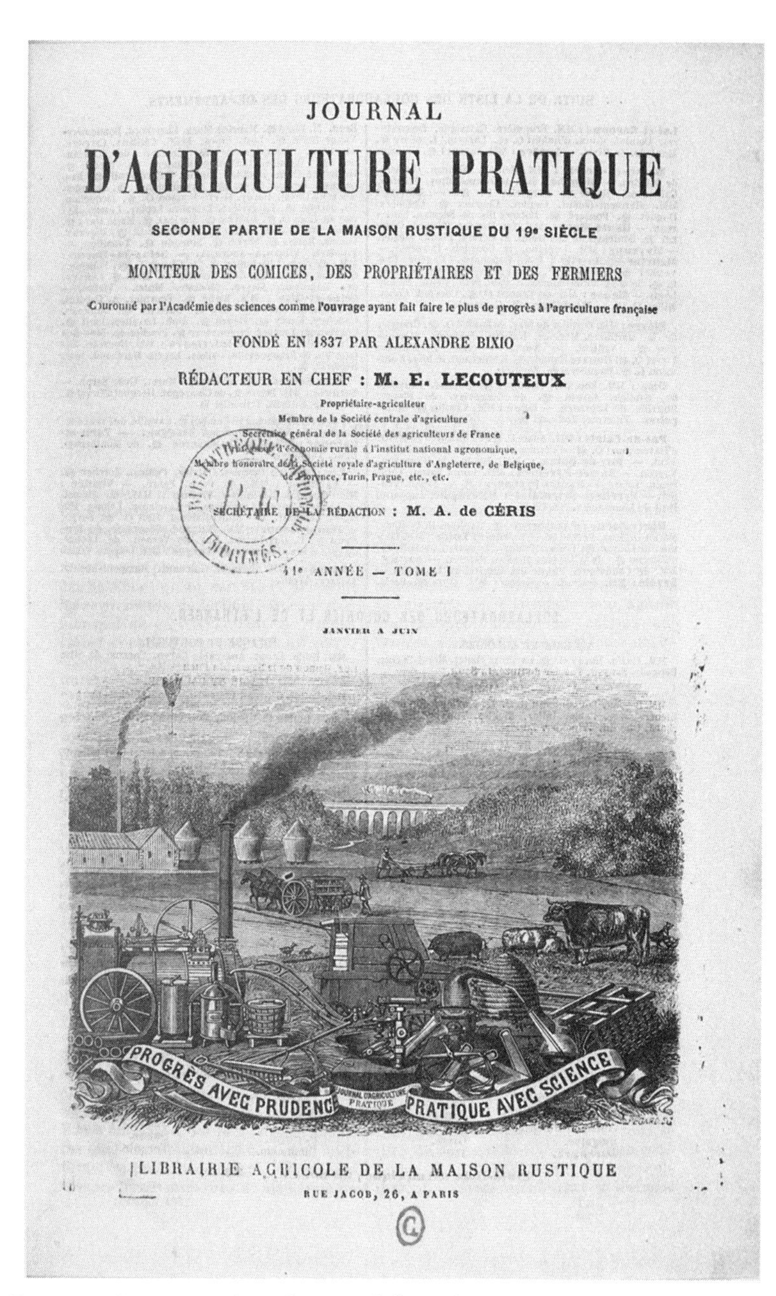
JOURNAL

D'AGRICULTURE PRATIQUE

SECONDE PARTIE DE LA MAISON RUSTIQUE DU 19e SIÈCLE

MONITEUR DES COMICES, DES PROPRIÉTAIRES ET DES FERMIERS

Couronné par l'Académie des sciences comme l'ouvrage ayant fait faire le plus de progrès à l'agriculture française

FONDÉ EN 1837 PAR ALEXANDRE BIXIO

RÉDACTEUR EN CHEF : M. E. LECOUTEUX

Propriétaire-agriculteur
Membre de la Société centrale d'agriculture
Secrétaire général de la Société des agriculteurs de France
Professeur d'économie rurale à l'Institut national agronomique,
Membre honoraire de la Société royale d'agriculture d'Angleterre, de Belgique,
de Florence, Turin, Prague, etc., etc.

SECRÉTAIRE DE LA RÉDACTION : M. A. de CÉRIS

41e ANNÉE — TOME I

JANVIER A JUIN

LIBRAIRIE AGRICOLE DE LA MAISON RUSTIQUE

RUE JACOB, 26, A PARIS

Figure 1.4. Cover image from the *Journal d'agriculture pratique* in 1878. In contrast to figure 1.3, this image, produced over thirty years later, features a modernized agricultural landscape transformed by steam power. Windmills have disappeared and healthy forests are abundant. The banner at the bottom of the image, summarizing the new age of scientific agriculture, reads: "Progress with Prudence, Practice with Science." Source: gallica.bnf.fr/BnF.

plow pulled by two oxen. A small, isolated village with a church tower appears behind him. The lands surrounding the village appear dry, with little vegetation. Far in the distance, three windmills are perched on hilltops, turning in the breeze.

After 1850, however, the arid environment featured on the journal's cover is replaced by a new and improved French agricultural landscape (see figure 1.4). In the foreground, the banner proclaiming "Progress with Prudence, Practice with Science" stretches below an assemblage of modern farm machinery. Behind the machines, a vision of abundance unfolds. Livestock appear healthy and robust, fields are fertile, and the presence of grain houses signals excess production. In the distance, a train, puffing smoke into the air and crossing an elevated bridge, transports the farmers' bounty from the countryside to the hungry residents of France's growing cities. The hillsides surrounding the railway bridge are completely covered with healthy, vibrant forests. For nineteenth-century agricultural reformers, the second image represented the fulfillment of the promise of science and a vision of modern progress rooted in national economic growth and prosperity. Not everyone in Provence, however, agreed that this modernizing attitude toward agricultural lands was a step in the right direction.

Breath of Provence: Refashioning the Mistral as Regional Patrimony

While French administrators, foresters, and agricultural experts believed that bringing order to unruly provincial environments was the key to civilizational progress, prominent members of Provence's literary and artistic community questioned the idea that taming the nation's diverse regional landscapes would necessarily lead to a brighter future. In 1854, a group of Provençal poets and writers founded the Félibrige, a regional association dedicated to defending the Provençal language and identity. While the regionalists' goal of preserving their provincial culture has been well documented, scholars have paid surprisingly little attention to the Félibrige's fascination with the Provençal climate.[69] In addition to situating themselves within a specific regional culture, members of the Félibrige movement rooted themselves in a place-specific climate system that they believed had a powerful influence over the Provençal body, mind, and spirit.

The Félibrige's interest in the bonds between traditional Provençal society and its surrounding natural environment has a long precedent in French history. For Jacobin revolutionaries seeking to bring the light of progress and civilization to the nation's "backward" provinces, the close relationship between the French peasantry and the natural world was perceived as a problem. In 1794, the rebel priest known as the Abbé Grégoire organized a survey of language and culture in the French provinces with the aim of identifying areas in the Republic that needed civilizing. Provincial government administrators wrote harshly about the rural communities under their control, describing large swaths of the French countryside as mired in "a sort of permanent Middle Ages."[70] Peasants, in their view, lacked any intellectual capacity, leading lives that were "purely physical."[71] A French peasant was "the unwilling intermediary between nature and civilization . . . a creature comparable to a domestic animal."[72] Unlike French, the "primitive" patois that rural people spoke was the unrefined byproduct of the land, the organic result of "an ecology, a geography, an anthropology, a history."[73] In order for France to come together as a unified nation, the Jacobins argued, the "natural" links between peasants and their heterogeneous regions had to be severed, replaced by a universal French civilization.

But at the same time that some French revolutionaries criticized rural France for its backwardness, other Enlightenment- and Revolutionary-era thinkers saw peasants' close ties to nature as a perfect and beautiful antidote to the inevitable corruption that civilization brought with it. Why tame nature, asked the Romantics, when there was so much to be revered about it? Admirers of Jean-Jacques Rousseau, the Romantics saw nature as a site of spiritual rejuvenation. Theirs was a vision of "a wild nature of passion and imagination as opposed to a scientific nature of decipherable logic and attainable order."[74] During the early nineteenth century, the Romantics inspired a widespread movement to preserve the French nation's wild areas from development. Primarily directed at the local level, these "preservationist" initiatives found common cause with regionalists seeking to stave off threats from modernizing government and commercial institutions headquartered in Paris.[75]

Provence became one of the leading centers for regional preservation and regionalist resistance in nineteenth-century France. Centered in Arles, the Félibrige movement united writers, clerics, scholars, and ethnographers together in a shared mission of preventing Provence's unique landscapes from disappearing. Promoting a type of exclusive territorial identity defined in terms of "blood and soil" and "the Provençal race," the Félibrige was hostile to "unrooted" people, including Jews and immigrants.[76] For this group of conservatives wary of outside influences and leftist politics, Provençal nature offered a reassuring sense of permanence. One of the oldest of the Félibres, the poet Antoine-Blaise Crousillat, wrote a poem that expressed the Provençal people's enduring human bond with the mistral, a force of nature that had been with them since time immemorial:

> Oh gardens of olive trees watered by the Durance!
> Fresh valleys, fertile slopes
> O resplendent fields of lower Provence,
> Why aren't you Paradise?
>
> It is the terrible and violent mistral, impetuous blower,
> That takes from you this honor,
> When, o my country, it pillages everything that is the pride of your soil
> With the breath of its furor . . .
>
> But, often, its rages are pointless against you;
> Often, lovingly mild,
> It courteously chases for you mosquitos, storms
> And it sends worry far away.
>
> If every good Provençal is the friend of joyfulness,
> To which he owes the mistral . . .
> What do you know, is it not thanks to a light mistral that my cheerful muse
> Owes its breath of inspiration?[77]

A wind that was simultaneously "terrible and violent," "lovingly mild," and "a breath of inspiration," the mistral symbolized for Crousillat an authentic Provence that was simple, violent, and alive.

Similar praise for the mistral could be found in the works of another celebrated Provençal regionalist, Frédéric Mistral (see figure 1.5). "Like the wind whose name becomes you, strong is the holy breath that inspires you," wrote Crousillat in a poem about his fellow Félibre. Born the son of a prosperous mas owner in the small Provençal town of Mallaine, Frédéric Mistral was an educated bourgeois who became passionate about defending Provence from the threats of modernity. His most famous work, an epic poem called *Mireille*, or *Mirèio*, which he published at age twenty-nine, in 1859, was dedicated to "the shepherds and the inhabitants of the mas." The poem tells the story of a young Provençal girl who falls in love with a farm boy and escapes from her coastal home in Saintes-Maries-de-la-Mer when her father refuses to let them marry. In narrating his heroine's journey through Provence—across the Camargue, the plains of the Crau, and the Rhône River—Mistral created

Figure 1.5. Photograph of Frédéric Mistral in 1909. Sharing his name with his native region's famous wind, the Nobel Prize winner was a leading advocate for protecting Provence's language and cultural traditions from industrial modernity and Parisian overreach. Source: gallica.bnf.fr/BnF.

a personal dialogue between Mireille and the natural world of his native region.

In Mireille's fantastical experience of the rural Provençal landscape, the mistral animates the region's vegetation, making it come alive. In one scene, Mireille is taking part in the festivities of St. John's Day

(June 24) when she encounters a group of reapers near the Mas des Grands Micocouliers. The "impetuous mistral" began to shuck the wheat and the stalks themselves began to cry out, asking to be harvested: "Master, they murmured, it is time! Watch how the *bise* leans us over—and sheds us and deflowers us!"[78] In Mireille's Provence, it was a higher power, not man, who controlled the wind: "God opens his hand; and the Mistral—with lightning and storm—leaves his hands like three eagles; from the deep sea, to its ravines, and its chasms, they go, eagerly."[79] The wild and disorderly mistral, in other words, was part of God's plan for Provence. Where the time-honored presence of the mistral brought life to the people, modernity brought death. Technological progress, in Mistral's view, was a tragedy, a "terribly fatal hearse, against which there is nothing to do or say."[80]

Subscribing to similar regionalist views as Mistral, Alphonse Daudet, the conservative French writer who became famous for his affectionate portraits of southern French life, mourned the disappearing bonds between French people and the natural world. One of his short stories, "Old Cornille's Secret," offers a moving parable about a Provençal landscape and a Provençal way of life that were vanishing. Inspired by the town of Fontvieille, Provence, where Daudet's wife spent her childhood, the story focused on the heartbreaking fate of an old windmill.

"There was once a flourishing flour-milling trade here," the narrator in Daudet's story began, "and farmers from ten leagues around would bring us their wheat to grind . . . Whichever way you looked, all you could see were sails turning in the mistral above the pine trees."[81] Then the windmills of Provence started to disappear:

> Unfortunately, some Frenchmen in Paris had the idea of setting up a steam-powered flour mill on the road to Tarascon. All shiny and new, it was, and so people got into the habit of sending their wheat to the steam mill, and soon the poor windmills stood idle. For a time, they tried to compete, but steam power was stronger and—alas!—one after another they were all forced to close. However much the mistral blew, the sails stood still.[82]

But amid the rapid demise of the old village economic order, there was one mill—and one miller—who refused to shut down. The defiant miller, a character called Old Cornille, had spent sixty years of his life working in the flour industry. When the steam mills arrived, Old Cornille was driven insane with sadness and despair. "Stay away from here," he warned his fellow villagers, "those crooks make their bread using steam, which was invented by the devil, whereas me, I work in harmony with the mistral winds, which are breathed by God himself."[83]

Remarkably, Old Cornille's mill seemed to be surviving despite the steady expansion of industrial milling. Every day, bags full of flour made their way down the hill from his mill. The villagers were baffled. How could the old windmill still be turning a profit? The narrator decided to investigate by pushing a ladder up against the mill's window to see what was going on inside. To his horror, he saw the old man filling the bags, not with flour, but with white plaster. Shocked, embarrassed, and ashamed, the narrator called the village together and, in an expression of Provençal regional pride, the members of the village loaded up their donkeys with wheat and marched up to Cornille's mill to have their grain processed in the old-fashioned way. The elderly man wept in happiness, and for a brief time the traditional wind-powered and community-centered economic order returned. In the long run, however, the march of technological change was too forceful to be stopped. Daudet's story ends with Old Cornille's death, and the death of his windmill alongside him.

But it was precisely when it ceased to be a functioning mill that Fontvieille's Moulin de Saint Pierre, otherwise known as Daudet's windmill, transformed into an icon of Provençal regional identity. The historic preservation of the shuttered mill was part of a broader museumification of the Provençal agricultural landscape. In late nineteenth-century tourist guidebooks, Daudet's mill was highlighted as a place where visitors could experience, and pay homage to, "authentic Provence." The mill's photographic image was reproduced tens of thousands of times on postcards that visitors could collect as souvenirs (see figure 0.5). Like the mas farmhouse and the Camargue hut, Daudet's windmill became an architectural memento of a disappearing way of life in Provence.[84] Frozen in time while France's future was being decided elsewhere, the old provincial buildings were part of a nostalgic landscape in which human bonds with nature were still intact. It was a landscape that French political conservatives praised as an "organic" antidote to the corrupt cosmopolitan world of cities, where nature, long forgotten, no longer informed the rhythms of daily life.

Conclusion

Neither fully natural nor entirely human made, Provence's transforming modern landscape embodied both the sculpting power of the mistral and the adaptive designs of the human imagination. During the early modern period, the changes that people made to Provençal territory were modest, involving the application of dry-agricultural techniques and the construction of wind-resistant peasant dwellings made from locally available materials. But the late eighteenth and nineteenth centuries unleashed a wave of dramatic modifications to the countryside. In addition to bringing "enlightenment" and "progress" to French politics and industry, modern French administrators sought to pull French weather out of its uncivilized past and into a well-regulated and prosperous future. The result was an aggressive state-sponsored landscaping project that dramatically altered Provence's windswept terrain through mass reforestation and windbreak plantings aimed at slowing the mistral's powerful and destructive gusts.

While it became the target of a national-scale plan to assert centralized environmental control over provincial territories, the mistral also became entangled in a middle-class-led movement for regionally situated landscape conservation. In exalting ordinary people from the countryside for communing with the mistral and lamenting the modern Frenchman and Frenchwoman's estrangement from the life-giving forces of nature, Provence's Félibrige movement promoted its own form of conservative environmentalism. Thanks to the Félibrige, the traditional wind-adapted buildings that had once dotted the Provençal countryside—the windmill, the mas, and the *cabane de Camargue*—transformed from utilitarian works of wind-adapted architecture into defunct yet powerful icons of an environmentally rooted regional identity.

Inland Provence's transformation from a traditional wind-centered agricultural society into an experimental landscape for state-directed scientific agriculture opened up lively public debates over the meaning, use, and preservation of regional nature in modern France. In the next chapter, we will explore how these same forces of modern "progress"—techno-optimism, large-scale capitalism, and nation-state power—clashed with wind-centered maritime communities in Provence situated along the Mediterranean coastline and the Rhône riverway. In places like Marseille and Arles, where ordinary people's livelihoods revolved around knowing the mistral and navigating unstable bodies of water in its presence, the arrival of steam power and modern industry marked the decline of vernacular sailing expertise. As communities of sailors became subsumed into industrial shipping and factory jobs, the localized weather knowledge they had once cultivated became the purview of wealthy new middle-class arrivals who flocked to Provence for the leisure sailing industry made possible by its unique regional windscape.

CHAPTER 2

The Lion's Roar

MEDITERRANEAN JOURNEYS WITH THE MISTRAL

> To have a port in the Wind means that no matter the difficulties that come your way, you are certain to overcome the situation before you—To head into the Wind, means that you will encounter a lot of obstacles in accomplishing your goals, or that the efforts that you make will bring about a good result—On the other hand, to have the Wind at your back signals that all possible difficulties will disappear magically, and that you will get whatever you wish.
>
> PIERRE-MARIE-JOSEPH DE BONNEFOUX AND ÉDOUARD PARIS, *Marine Dictionary for Sail and Steam*, 1859

On the days when the mistral is blowing, the Mediterranean coastline of Provence visibly changes. The water turns what Émile Zola describes as a "somber blue"[1] and the wind's powerful force repels the sea's waves from the coast, sending them backward, in what appears to be a reverse tide. When the mistral pushes surface water away from the shores, it allows cold water to move up from the deep. The wind-driven cycling of seawater can cause its temperatures to plummet ten degrees Celsius in the span of a single day, a phenomenon with which Mediterranean bathers are painfully familiar.[2] Farther out from the coast, the mistral generates even more dramatic wave activity, transforming the open sea into a choppy expanse of whitecaps. Provençal locals call it a *mer moutonneuse*: a sea that is so full of foam and froth that it resembles a herd of woolly sheep. Sailors have used other metaphors, likening a mistral-blown sea to "a plain covered with snow" and a sea "strewn with ashes."[3] And then there is the sound. Legend has it that the Gulf of Lion, the body of water surrounding coastal Provence, got its name from sailors who heard the terrifying "lion's roar" of the mistral as they struggled to keep their vessels upright.[4]

The mistral originates over land, but its final destination is the sea, with nearly half of its geographic range extending over the Mediterranean. For the coastal communities in Provence whose traditional livelihoods revolved around maritime commerce, transportation, and fishing, knowing how to navigate a windswept seascape was imperative for their physical and economic survival. In contrast to farmers and vintners who could rely on windbreaks and other types of landscape modifications to block the mistral, Provençal mariners could do nothing to alter or "improve" their watery geography of labor.[5] Rather, sailors and fishers leaned on their physical skills and their vernacular knowledge of the regional atmosphere to work safely and effectively in the wind.

Operating their vessels in a dynamic sea environment, Provençal sailors' bodies were closely attuned to nature; they sensed the first signs of changing weather and became experts on coastal geography both above and below the water's surface. Throughout the early modern period, the mistral's hazardous seasonal gusts guided mariners' sense of time, setting the rhythm of their annual work schedules. It informed the sailing routes they followed and the safe harbors where they sheltered. The mistral influenced their choice of sailing vessels, generally small craft such as lighters (*allèges*) and tartans (*tartanes*) that could maneuver well in windy conditions.

But new economic pressures challenged Provençal mariners' close-knit working relationships with the wind in the nineteenth century, when the proliferation of steam power meant that seafaring vessels no longer depended on organic forces to move. Instead of working in tandem with the wind, capturing its natural source of energy with handmade sails, steam-powered boats enabled mariners to plow directly into the wind—even a powerful mistral—without the time-consuming practice of tacking. Sailors from coastal Provence who were once experts in manipulating topsails and adjusting riggings morphed into industrial laborers—technicians, mechanics, and engineers—who fed, fixed, and operated the steam engines onboard in a seasonless manner resembling factory workers on land. By the end of the nineteenth century, time condensed and space shrank as mighty steamers journeyed from Marseille to Italy or across the sea to Tunisia in less than half the time as they had before.[6]

While the nineteenth-century industrial age weakened maritime laborers' nuanced working relationships with their windswept environs, it fostered a growing interest in sailing among members of the burgeoning French middle class. New forms of pleasure craft such as yachts offered bourgeois tourists from northern industrial cities the opportunity to experience the coastal environment of Mediterranean France in a direct, embodied manner. Thanks to the rise of regional tourism and leisure sailing, Provence's windy atmosphere continued to play a major role in the region's economic livelihood long after the collapse of its sail-based commercial fishing and transportation industries.

How Mediterranean Sailors Visualized the Wind

Sailing the Mediterranean Sea was, for most of human history, a risky proposition that struck fear in human beings. A seemingly bottomless abyss whose opaqueness masked a grotesque realm below, the sea was a place that traditionally conjured up repulsive images of sea monsters and demons.[7] For sailors, fear of what lay beneath the depths of the sea was compounded by the violent and capricious winds that blew across its surface. According to Fernand Braudel, the ancient societies of the Mediterranean Basin were "almost as daunted by the sea as later generations were by the sky," and "did not risk sailing the Mediterranean until the twelfth and eleventh centuries, BC, at the earliest, and more likely in the sixth and fifth."[8] The Mediterranean's slow transformation from a place of fear into a legible, known, and navigable body of water was made possible by the spread of new navigational tools, particularly maps and compasses, that helped sailors to visualize both the sea itself and the invisible winds that passed over its surface.

The first "map" of Mediterranean winds was not a conventional map at all, but a directional arrangement of winds called a wind rose, or a *rose des vents*. Originating in ancient Greece, wind roses offered simple schema for ordering and rendering visible the names and cardinal directions of prevailing winds. Over time, they changed form and purpose as they became entangled with new cultural geographies.[9] During the period of the Crusades, crosses often appeared on Mediterranean wind roses next to the designation for the east wind. Later, artists blended racialized geographies into wind imagery. In a decorative seventeenth-century wind rose printed in the *Neptune françois*, a lavish maritime atlas commissioned by Louis XIV and published in 1693, the circle of Mediterranean winds is surrounded by four faces, or wind heads, blowing onto the earth (see figure 2.1). While the oldest, most distinguished-looking face is a light-skinned bearded man representing the north wind (Septentrion), the east wind (Orient) stands out as the only wind head to take the form of a dark-skinned cherub. The wind rose's title itself underscores the role that cultural bias played in the practice of mapping and naming winds. Entitled "Compass of Winds, or Their Ancient and Modern Names Divided into Six Circles According to the Principal Nations of Europe," the image reveals the dual

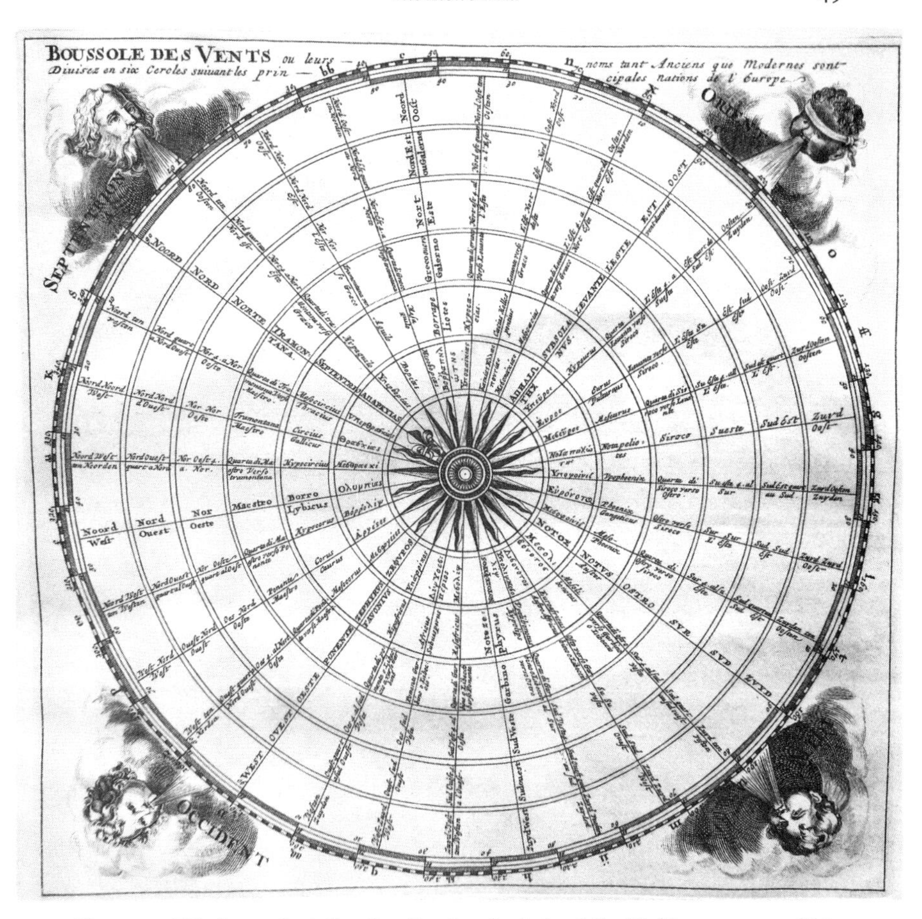

Figure 2.1. Wind rose depicting the directional winds of the Mediterranean world in different languages. In each corner of the image, wind heads blow air from the heavens down upon the earth. "Boussole des Vents ou leurs noms tant Anciens que Modernes sont Divisez en six Cercles suivant les principales nations de l'Europe," in Nicolas Perrot d'Ablancourt, Giovanni Domenico Cassini, and Charles Pêne, *Le Neptune François* (Paris: Hubert Jaillot, 1693). Courtesy of the Newberry Library.

nature of winds as palpable physical forces and reflections of cultural belief systems.

Beginning in the thirteenth century, wind roses were commonly found on a new type of maritime map called a portolan chart. In contrast to medieval *mappae mundi*, which created a blended view of Mediterranean geography based on both biblical scripture and real-world geography, portolan charts were largely secular documents directed toward the practical needs of Mediterranean navigators and merchants. Originally hand painted on animal skins, by the fifteenth century, portolan

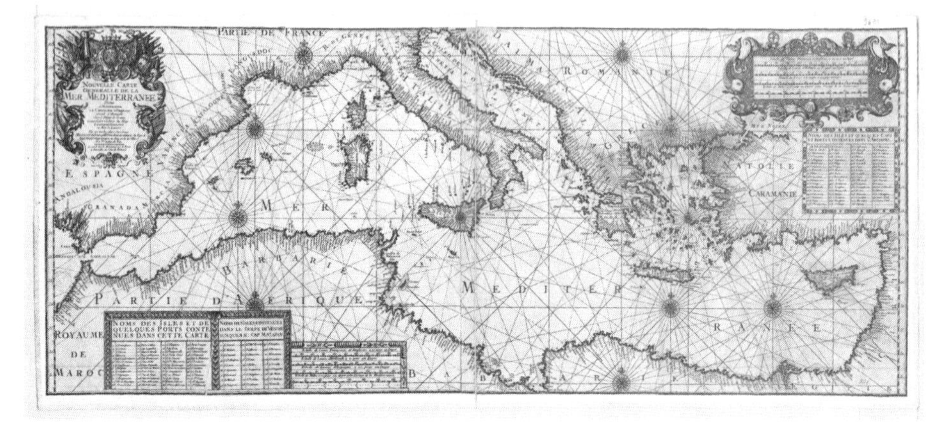

Figure 2.2. For the hundreds of Mediterranean coastal trading communities found on this portolan chart, winds played an essential role in everyday life and labor. Sailors used compasses and rhumb lines—the crisscrossing lines that cover the chart—to navigate from one part of the Mediterranean Basin to another. Henri Michelot and Laurent Brémond, *Nouvelle carte generalle de la Mer Mediterranée* (Marseille: L. Brémond, ca. 1700–1799). Source: gallica.bnf.fr/BnF.

charts evolved from manuscript to printed versions, making them accessible through the commercial market to average sailors. In their visual form, portolan charts promoted a borderless, interconnected vision of Mediterranean geography that pulled coastal communities together into a web of commercial relationships.

The eighteenth-century *Nouvelle carte generalle de la Mer Mediterranée*, produced by Marseille-based mapmakers Henri Michelot and Laurent Brémond, exemplifies the visual style of a modern portolan chart (see figure 2.2). Loxodromic (or rhumb) lines, spinning out like bicycle spokes from tiny wind roses pointing north, cover the map. Each line represents a pathway that a ship could use to orient itself out of sight of land with the help of a mariner's compass, a magnetic device that entered European sailing culture from Arab traders. The portolan chart's emphasis on the Mediterranean Sea as a cohesive, legible, and navigable space is made all the more salient by Michelot and Brémond's lack of attention to detail in the European and North African landmasses, which blend together in a boundless expanse of insignificant white space. The only parts of land that the mapmakers represented in any detail on their map were coastal regions—such as Provence, Languedoc, and Catalonia—and hundreds of coastal ports, which were labeled in tiny letters along the littoral outlines of the Mediterranean Basin. It was not uncommon to find sixteenth- and seventeenth-century wind roses that appealed to audiences from several Mediterranean cultures, such as this

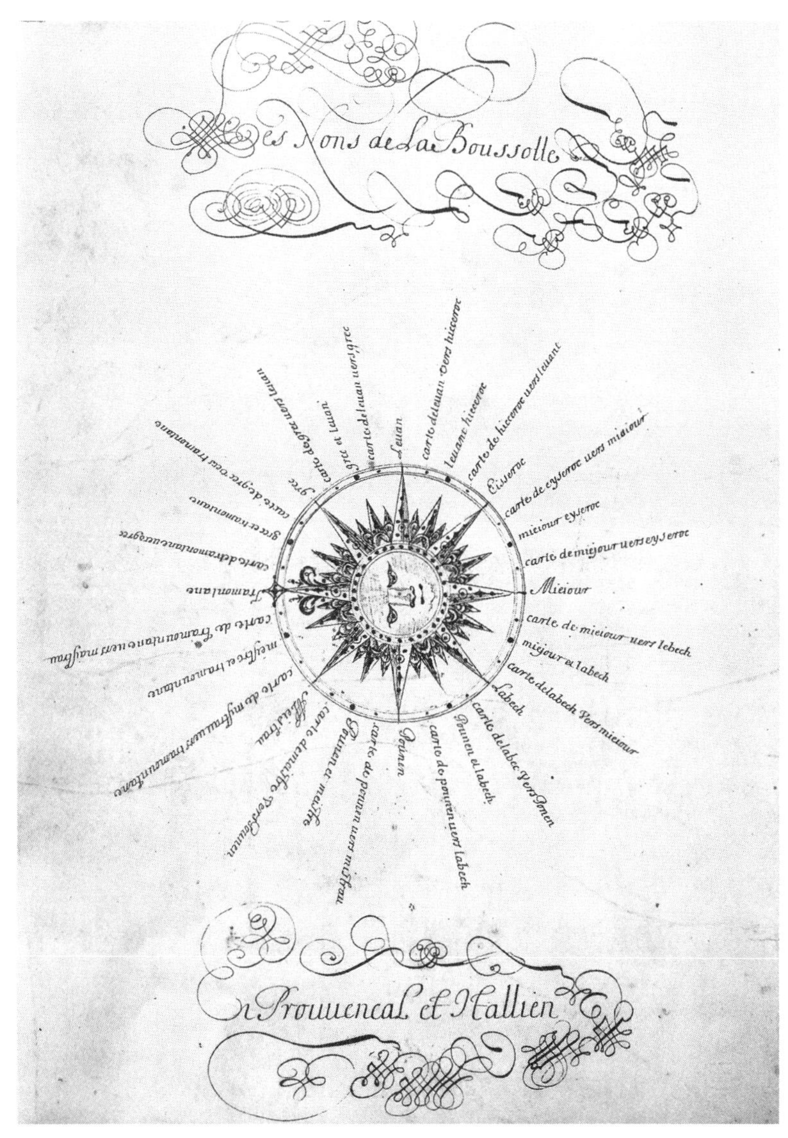

Figure 2.3. This seventeenth-century wind rose in the “Prouvencal” and “Itallien” languages revealed how neighboring cultures in the Mediterranean Basin shared their working knowledge of winds. “Les Noms de la Boussole en Prouvençal et l’Itallien,” in François Ollive, *Atlas-portulan du bassin méditerranéen*, 1646. © Bibliothèque de Marseille, Fonds patrimoniaux: Ms2125_f_6.

Figure 2.4. City map of Marseille from 1640. The wind's imprint is visible across Marseille's urban space and in its surrounding natural environment. Note the presence of wind in the mills standing atop the hills of the Vieux Panier district, in the sails of boats, and in the undulating movement of the Mediterranean waves. Image from Georg Braun and Franz Hogenberg, *Civitates Orbis Terrarum*, 1640. Source: gallica.bnf.fr/BnF.

expressive wind rose that provided "The Names of the Compass in Provençal and Italian" (see figure 2.3).[10]

Many of the seventeenth- and eighteenth-century portolan charts used by Mediterranean sailors were designed and printed in Marseille, a Provençal city closely intertwined with both the wind and the sea. A 1640 map of Marseille printed in the first-ever atlas of world cities, the *Civitates Orbis Terrarum*, pictured the tightly walled and gated urban space in its dynamic windblown setting (see figure 2.4). To the left, the Mediterranean appears alive with the wind's motion; the viewer can spot some boats out in the choppy sea while others are nestled safely in the port. Even the hills surrounding Marseille appear to undulate with movement. Inside the urban settlement, rows of windmills can be spotted on the hilltops where their sails caught the gusts of air that passed freely over the city's defensive walls. Known as the Butte des Moulins, this hilly area is where inhabitants of Marseille milled bread so that residents in the city had a stable source of food during a siege. Another windmill appears on a coastal island offshore. The atlas image offered a portrait of Provence's largest city as a place of encounter between a human-generated urban geography and a natural geography fashioned by water and wind. Within the city's walls, meanwhile, generations of sailors were learning to navigate through the complexities of both.

The Art of Navigation: Sailors' Embodied Knowledge of the Wind

To grasp how sailors journeyed on the Mediterranean Sea in the age of sail, it is helpful to think of navigation as less of a science and more of an art form that, on any given day, could include unfurling or furling sails, avoiding underwater rocks, or retreating to safe harbor. The skills and tools that navigators incorporated into their sailing art were extensive. While some forms of sailing know-how could be transmitted from generation to generation in paper form—through maps, charts, and pilot books—other forms of knowledge had to be acquired firsthand. Learning how to read a changing maritime environment and make sense of its visual and sensory cues took years of training and experience. By the sixteenth century, Provençal sailors' growing confidence in their ability to orient themselves at sea and to harness the wind in their sails created an economic environment in which trade among Mediterranean cultures flourished.[11]

In Provence, mariners planned their commercial, fishing, and diplomatic sea journeys around seasonal variations in prevailing winds. From late October until March, when the mistral was most severe, sailors eased up on their commercial routes along the coast of Provence. If they could not avoid sailing into the wind—either to make an important trading deal or to sign a diplomatic treaty—their only choice was to proceed by tacking, a technique that involved pivoting the boat in zigzag fashion, moving sails or oars to the left and right, in order to make forward progress. Smart boat design could also help to alleviate wind resistance.[12] But even with the best-designed hull, tacking could add hours, if not days, to a sea journey. One historian estimates that Mediterranean voyages made against prevailing winds could take twice as long as a typical journey.[13]

One of the most profitable and efficient trading itineraries in Provence, the sailing route from Arles to Marseille, set its clock directly

Figure 2.5. A model of an *allège* (lighter) created by Captain Calment at the end of the eighteenth century and completed by Henri Boudignion in the twentieth century. *Allèges* were long boats with flat bottoms, measuring twenty meters long, with a large Latin sail and a jib in the front called a *polacre*. Adapted for commercial navigation on the lower Rhône River, these boats were created to transport salt and other merchandise. In the nineteenth century, many were converted into barges before disappearing from use. © S. Normand - Cd13, Museon Arlaten–musée de Provence.

to the behavior of winds. Patrons of the shipping industry in Arles, who had access to goods such as wheat, wine, and salt, would hire sailors to carry their products downriver to Marseille and Toulon on small sailboats called *allèges*, or lighters (see figure 2.5). The flat-bottomed boats were designed for loading and unloading goods efficiently and could be effectively maneuvered in challenging weather. Sailors from Arles, up through the mid-nineteenth century, would time their journeys down the Rhône River around the directional force of regional winds. "The mistral, or the winds from the sector of the north or northwest were essential for accomplishing the southern route," explain historians Patricia Payn-Echalier and Philippe Rigaud. "To the contrary, those from the sector of the south-west were necessary to carrying out the return trip." This pattern resulted in a rush of boats traveling upriver to Arles when the winds shifted: "A certain number of boats carried out their commercial operations in different ports along the coast of Provence and waited to leave until the first wind from the southeast . . . It was not unusual to see forty or fifty boats going back up the Rhône and returning to Arles in a single day."[14] An intriguing side effect of the back-and-forth trading pattern was the navigational opportunities that it offered for women and girls who were often tasked with sailing the empty boats back up the Rhône River to Arles.[15]

In addition to working in harmony with regional wind patterns, Provençal navigators exposed to the open Mediterranean Sea would choose shipping routes that provided the greatest amount of topographical protection from the wind. "A coastline always in sight is the navigator's best aid and surest compass," explains Braudel. "Even a low-lying coast is a protection against the sudden and violent Mediterranean winds, especially the offshore winds. When the *mistral* blows in the Gulf of Lions, the best course even today is to keep close to the coast and use the narrow strip of calmer water near the shore."[16] Specifically, mariners used "trunk routes": sea lanes that followed the outline of the coasts from Barcelona to Marseille and along Italy's Ligurian coast. To reach Africa, mariners would either hug the west coasts of Corsica and Sardinia or go down the coast of Italy.[17] This navigational technique was so widespread that it earned its own name—*cabotage*—and the people who took part in Mediterranean coastal navigation were called *caboteurs*, or coastal pilots. As late as the 1830s, cabotage accounted for nearly half (47 percent) of the total volume of Marseille's commerce.[18]

"In its most widespread meaning," explained one nineteenth-century sailing guide, "the coastal pilot is synonymous with the art of navigation. The coastal pilot is . . . the man who knows the coasts, their appearance, their angles and their bearings, the measurements and quality of their depths, the banks, the dangers and the reefs, the different passages that should be followed to arrive safely . . . The knowledge of the coastal pilot is purely local and practical."[19] A logbook left behind by one *caboteur* from Arles named Pierre Giot offers a glimpse of the wind's prominent role in his daily work, movements, and decision-making:

> In the morning, there appeared a northwesterly wind that was almost calm . . . At 8 o'clock, the wind changed to a west-northwest, deep chill [*grand frais*] and rough sea [*grosse mer*] to the west. Performed

> several sideways moves to make it to Bandol and Senary . . . The wind became increasingly strong; we arrived and dropped anchor at the same place . . . A little later, a tartan came and dropped anchor. During the day, the wind strengthened and, in the night, a storm [*fortune*].[20]

Giot, like many *caboteurs* of his generation, came from a simple background: his father was a barge worker and his great-grandfather a valet at a mas. Coastal piloting was a skill that he learned from his father, who took him along on barge trips as a ship's boy or *mousse* (according to French law, a boy could begin working as an apprentice sailor at the age of twelve). By age sixteen, he became a novice; at age eighteen, he became a sailor (*matelot*); and at age twenty-eight, in 1775, he was promoted to captain. Acquiring no formal schooling, Giot learned everything he knew about sailing on the job.

Like those of most ordinary French mariners of the late eighteenth and early nineteenth centuries, Giot's navigational techniques were rooted in his firsthand, embodied knowledge of nature. His primary language was Provençal, and his logbook, a blend of French with some Provençal words phonetically sounded out, reflected his regionally situated culture and training. His logbook's attention to shifting meteorological conditions and physical changes to the Rhône riverbed demonstrated his keen awareness of environmental threats.[21] As vigilant as he was in observing the sky, the sea, and the coastline, however, Giot was always prepared for his boat to encounter trouble. Onboard he carried a wide range of repair tools, including nails, boards, sails, thread to sew the sails, extra ropes, supplemental materials for masts, and tar and caulking materials for sealing leaks.[22] In addition, at his disposal were different types of skilled maneuvers that he had learned over the years when facing unfavorable winds, including changing, furling, and unfurling sails to change course quickly and efficiently.

Evidence from Provençal popular culture shows that ordinary sailors like Giot would have supplemented their own embodied knowledge of their regional atmosphere with animal-bodies-turned-barometers that signaled changes in the weather. Known as *devino-vènts*, these dried bodies—plucked right out of the nearby landscape—were hung from the ceilings or roofs of sailors' homes, where their appearance was thought to change in tandem with small shifts in atmospheric conditions.[23] The Museon Arlaten, in Arles, houses an example of a *devino-vènt* in the form of a dried-out *malarmat* (armored sea robin), a local fish with two little horns on its head, which was suspended from a roof joist in a Camargue hut to indicate the direction of the wind and to announce

Figure 2.6. An example of a *devino-vènt* in the form of a dried fish called a *malarmat*, or an armored sea robin. Suspended from roof joists in Camargue huts, *devino-vènts* indicated the direction of the wind and predicted weather conditions through changes in their physical condition. © S. Normand - Cd13, Museon Arlaten–musée de Provence.

good weather or rain through changes to its physical constitution (see figure 2.6). The museum also houses a desiccated kingfisher's body that was suspended by the neck from the ceiling of a home. The bird's dried-out body was thought to possess material properties that made it sensitive to climatic variations. While these "bottom-up" weather tools have now passed into the realm of regional folk knowledge—and are framed as such at places like the Museon Arlaten—it is important to situate these handmade *devino-vènts* within a Provençal culture that conceived of the atmosphere as something that was physically embodied. By observing how the suspended, dried-out animal bodies interacted materially with the atmospheric conditions around them, ordinary Provençal people were making a larger claim that their regional climate and their material world were closely intertwined with each other.

Besides using locally sourced weather-measuring tools, *caboteurs* like Pierre Giot who wanted to sail to a new part of the Mediterranean coast, or to supplement their own knowledge of familiar terrain, also turned to printed sailing manuals called pilot books. Mass produced for sailors engaged in local commerce—who typically steered small sailing craft (boats like lighters or tartans, which carried only five to eight men)—pilot books offered precise visual and written descriptions of specific ports (and their dangers), helping readers to perform their delicate dance with the shore. In his pilot book published in 1805, for example, the experienced coastal pilot Henri Michelot assured his readers that his was a guide made "by a sailor and for sailors," and that his observations were all made directly by him, on site, unlike several Dutch pilot books that contained no experiential knowledge.[24]

When approaching the Port of Marseille, for example, Michelot explained, the trick was to drop anchor in a place that was safe from land winds (the mistral) but was also protected from winds to the south or southwest. If you were sailing in from the west and could not reach the Port of Marseille, he recommended seeking shelter in a nearby calanque, a natural limestone inlet. One could also hide from the wind in a particular spot on the safe side of the island of Ratonneau, not far from Marseille's harbor. In addition to alerting the reader to the location of these natural geological protections, Michelot pointed to human-made alterations to the coast that provided refuge from the wind, including recently built stone jetties, docks, and harbors.[25]

Even with years of sailing experience and pilot books onboard, exposure to the elements took a toll on the bodies of Provence's *caboteurs*. Pulling on rigging and adjusting sails required the sailors to exert themselves against the physical power of wind and waves. In theorizing the relationship between labor and nature, Karl Marx argued that the laborer "opposes himself to Nature as one of her own forces, setting in motion arms and legs, head and hands, the natural forces of his body, in order to appropriate Nature's productions in a form adapted to his own wants."[26] The sailboat was, from this point of view, "an instrument of labor . . . which the laborer interposes between himself and the subject of his labor, and which serves as the conductor of his activity."[27]

Indeed, French ethnographers and travel writers of the eighteenth and nineteenth centuries often noted the wind's long-term effects on the bodies of people who labored on the Mediterranean Sea. "The powerful winds that blew along the shore put the human fiber to the test . . . Consequently, men's skin became thicker, darker, drier . . . The inhabitants of the Provence coast [were seen] as a race toughened by the wind and the waves."[28] Studies on Provençal fishers, who relied primarily on small coastal craft, often complimented the men on the alertness of their bodies. "The body, the limbs of the fisherman must be both unconstrained and robust," explained nineteenth-century marine expert Sabin Berthelot. "His spirit and his eyes, always vigilant, always open, are continuously active."[29]

Belief in the human body's vulnerability to invisible forces of nature was also evident in emblems of religious conviction in Provence. While the Catholic Church as an institution weakened in France after the French Revolution, many people in Provence maintained a belief in God's power to control the behavior of the weather. After it was erected on an imposing hilltop overlooking the Old Port of Marseille in 1864, Notre-Dame de la Garde Basilica (known by locals as *la Bonne Mère*) soon became filled with hundreds of ex-votos gifted by laypeople. These humble devotional paintings were commissioned by members of Marseille's maritime community who wanted to make a public gesture of

Figure 2.7. Ex-voto of the boat *Le Benjamin* made by Captain Étienne Gombert in 1829. While sailing near Cap Couronne, to the west of Marseille, the captain was struck by a gust of the mistral, which destroyed his large sail and forced him to take refuge in his vessel's hull for an entire night. He eventually made it to safety at the Port of Agde the following day. Photograph on baryte paper of a graphic image by Captain Gombert. © Cd13, Museon Arlaten–musée de Provence.

gratitude to God for altering the course of a storm and saving them from a shipwreck. One of these ex-votos, created by Captain Étienne Gombert, pictures his boat, *Le Benjamin,* struggling on rough seas during a voyage in 1829 (see figure 2.7). While passing around the Cap Couronne, to the west of Marseille, his boat encountered a powerful gust of the mistral that ripped away its large sail. The captain was forced to spend the night in the boat's hull before managing to reach the Port of Agde, much farther to the west, the following night. In the upper-left-hand corner of the ex-voto, appearing in a swirl of dark storm clouds, are the holy figures of the Virgin Mary and the Christ child, whom Captain Gombert credits with saving his life.

Yet even as religious ex-votos commemorating near-death experiences at sea reinforced a widespread narrative of human fragility in the face of nature, a technological revolution was already underway in the nineteenth century to replace sails with a more secure motive power on the Mediterranean Sea. This new form of power promised to free sea captains, *caboteurs*, and other maritime laborers from their dependency on the unruly behavior of their regional atmosphere, enabling them to journey into a dangerous mistral without fear of losing their lives.

The Changing Ecologies of Boat Design in the Nineteenth Century

One of the most extensive visual archives of late eighteenth- and nineteenth-century boat portraiture—from grand ships to humble fishing vessels—can be found in the work of Marseille-based painter Antoine Roux. Growing up smelling the salt air and hearing shouts of longshoremen from his father's hydrography and naval-goods store on a quay along Marseille's Old Port, the young Roux passed his time observing and drawing the astonishing variety of sailing ships that floated before him: the giant brigs and the three-masted commercial ships of different nationalities, the Scandinavian and Hanseatic fluyts, the little sailing craft used for local commerce like lighters and tartans, the pinks from Genoa, and the American schooners. Before long, the captains, owners, and sailors who frequented his father's shop took note of Roux's talents and began commissioning him to make portraits of their ships.[30] For these working seamen, their ships were intimate dwellings—not unlike the vernacular houses pictured in chapter 1—and their desire to own portraits of their boats reflected their sense of attachment to their places of labor.

Roux's boat portraits were especially popular because he liked to capture sailing vessels that were actively engaging with their windy environment rather than standing still (see plate 3). He preferred to depict boats from a profile view, in the middle of a maneuver, their sails full of air, with a lively sea and sky accented with bright colors. But there was a strange irony to Roux's artistic oeuvre. His paintings celebrated the delight of sail-driven boats—their Romantic flair and their elegant dance with the natural elements—at precisely the historical moment when a technological revolution was about to render them obsolete.

The transition from wind- to steam-powered boats was part of a larger shift in nineteenth-century Europe, from what E. A. Wrigley calls an "organic economy" to an "industrial" one.[31] Organic economies, which had dominated Europe for much of its history, were powered by natural sources of energy like the sun, wind, and water. Transportation was fueled "by animal muscle on land and by wind at sea."[32] The invention and widespread dissemination of the steam engine offered Europeans a new source of unlimited mechanical energy that completely altered the possibilities for boat design. While a sailboat's speed was always limited by the direction and strength of the wind, boats powered by steam could overpower the wind's force and travel much more quickly and efficiently. Rather than figuring out how to sail *with* the wind, nineteenth-century navigators were intent on using mechanical energy to move *against* and *through* the wind.

In her study on population movements in the nineteenth century, Julia Clancy-Smith points to the advent of steam power as a key factor in accelerating the mobility of passengers and goods across the Mediterranean Sea. "Before the advent of steam," she writes, "passengers experienced delays, discomfort and great uncertainty" during their journeys from Europe to North Africa.[33] Beginning in the 1840s, French and British steamships began servicing a number of Mediterranean ports, reducing the duration of trips from weeks to days and even hours. By 1877, Tunis was a mere thirty-eight hours from Marseille aboard the fastest steamer.[34] In addition to facilitating the movement of thousands of entrepreneurs, traders, tourists, and missionaries between Europe and North Africa, steamers also made possible efficient diplomatic and postal communication between European metropoles and their colonies.

Sailors and ship manufacturers based in nineteenth-century Marseille, the largest maritime port in France and the fifth-largest port in the world, played a key role in establishing this new Mediterranean steamship regime.[35] The first steam-powered ship, a small Neapolitan craft, pulled into the Port of Marseille in 1818, and not long after, the Bazin brothers and François Fraissinet founded Marseillais steamship companies in 1831 and 1841, respectively.[36] Sailing routes that were once constrained and slowed by local winds, especially the mistral, were liberated by the flexibility and force of steam power. In his essay on French industrial progress, Michel Chevalier celebrated the economic boom that steam power had brought to the Port of Marseille:

> Look at the superiority of steamboats over sailboats; up until recently, coastal piloting between Marseille, on the one hand, and Cette and Agde, on the other, was done by *caboteurs* who, including the time it took to load and unload, only made an average of eleven or twelve trips per year. Steamboats can do it 100–115 times, or ten times as many. This speediness gives them the means to compensate for the expenses of their machines, and because business demands the same prices as they do for sail cabotage, the steamboats are able to provide entrepreneurs dividends of 25 or 30 percent.[37]

The logic of nineteenth-century capitalism, in other words, suggested that European nation-states should choose steam over wind if they wanted to speed up their supply chains and promote economic growth.

At the same time, French industrialists had to grapple with the fact that wind, though inefficient, was still "a motive force that nature provides for free."[38] Should French shipping companies really abandon the wind, an abundant no-cost source of energy, altogether? To answer this question, maritime engineers employed mathematical formulas to calculate the impacts of both favorable and unfavorable winds on steam-powered boats. Favorable winds, they determined, could enable a steam-powered boat equipped with sails to "reach and sometimes even surpass the maximum speed that it could attain from the action of a machine in calm weather."[39] A steamship captain could thus turn off his vessel's boilers and save precious amounts of coal while nature's gusts propelled them forward with sails, free of charge. But French naval architects had to be cautious in balancing the weight of sails with the weight of a steam-powered engine. "It impossible to give a steamship sails that are as large as boats propelled by the wind . . . The weight of the masts and the machines exhaust the hull," explained one naval engineer.[40] In the end, the nineteenth-century debate over which worked more efficiently—sails or steam—produced a significant number of curious-looking hybrids incorporating both sailing mast and smokestack that reflected a moment of transition between organic and industrial energy systems (see figure 2.8).

The advent of these hybrid steam-and-sail-powered boats fundamentally changed maritime labor in Provence. For centuries, wind-powered transport had sustained entire industries of skilled workers in Marseille. The work of sailmaking—cutting, sewing, and putting sails into a working state—was carried out by skilled artisans called *voiliers*. In the mid-nineteenth century, three factories in Marseille still specialized in cotton sailmaking.[41] The ropes used for riggings were produced by workers called *cordiers*. Marseille's most famous street, La Canebière, was named for the cannabis (hemp) that *cordiers* braided into thick ropes to use for rigging.[42] Shipbuilding itself, moreover, was an industry known for its artisanal roots. According to William H. Sewell, the workers involved in the fabrication of sailing ships in Marseille—ranging from caulk makers to marine carpenters—were members of skilled trades passed down from their fathers.[43] It is rare to find a blueprint for a ship built in Marseille, because most were constructed from models owned by a workshop master and then memorized by skilled workers.[44]

FRANCE.

DÉPARTEMENT DES BOUCHES-DU-RHONE.

PORT DE MARSEILLE

BATEAUX A VAPEUR.

PERMIS DE NAVIGATION

Délivré le au Bateau à Vapeur l'Aigle No. 1 Commandé par le Capitaine Pignon

Premier Mécanicien Réuel.

Nous,
Préfet du département des Bouches-du-Rhône,
Vu la demande, en date du du sieur Duqué Camille armateur, tendant à obtenir un Permis de Navigation pour le Bateau à Vapeur dit l'Aigle No. 1, ensemble les pièces annexées à ladite demande;
Vu le procès-verbal de visite du 2 Avril 1855, de la Commission de Surveillance des Bateaux à Vapeur du port d'Arles
Vu l'ordonnance royale du 17 janvier 1846, et l'instruction ministérielle du 6 juin de la même année;

ARRÊTONS:

ARTICLE PREMIER.

Le sieur Duqué Camille est autorisé, sous les conditions ci-après, à faire naviguer sur la Méditerranée, entre Marseille et Arles, en touchant à, le paquebot à vapeur dit l'Aigle No. 1, de la force de quatre vingt chevaux, jaugeant tonneaux, ayant une longueur de entre l'étrave et l'étambot, une largeur de à l'avant des tambours, et un tirant d'eau de un mètre en pleine charge.

ARTICLE 2.

Ce paquebot, destiné à transporter des marchandises, ne pourra recevoir plus de tonnes de chargement.

ARTICLE 3.

La ligne de flottaison sera tracée d'une manière fort apparente à l'extérieur du bateau

ARTICLE 8.

Le paquebot sera pourvu de deux embarcations, de trois ancres, d'une cloche, d'une bouée de sauvetage, de manomètres et tubes indicateurs de rechange, et des cartes et instruments nautiques nécessaires.

ARTICLE 9.

L'équipage sera composé au moins de huit personnes, savoir: un Capitaine

Il y aura en outre, pour le service de la machine, un mécanicien, un sous-mécanicien et deux chauffeurs et charbonniers.

ARTICLE 10.

Il sera tenu, à bord du bateau, un registre dont toutes les pages seront cotées et paraphées par M. le Maire d'Arles, et sur lequel les passagers et le capitaine pourront consigner leurs observations.

ARTICLE 11.

Dans chaque salle où se tiennent les passagers, il sera affiché une copie du présent Permis de navigation, ainsi qu'un tableau indiquant la durée moyenne des voyages et des relâches, le tarif des places, le nombre maximum des passagers, et la faculté qu'ils ont de consigner leurs observations sur le registre ouvert à cet effet.

ARTICLE 12.

L'instruction pratique du 5 juin 1846 sera affichée dans le local des machines.

Figure 2.8. As boat manufacturers recognized the usefulness of steam power, cities like Marseille had to set up new inspection procedures to ensure their safety. "Permis de Navigation, Port de Marseille," 1855. Archives départementales des Bouches-du-Rhône, VI S 1/1.

In terms of the types of workers involved in their operation, steamers (*paquebots*), in contrast to sail-powered ships, needed men to feed the boiler with coal and mechanics to keep the machines in working order. Like railway workers, steamship workers faced serious risks on the job. Steamboats were susceptible to the same terrifying explosions as land-based forms of steam transport.[45] Recognizing the potential hazards of steamship transportation, the Commission for the Regulation of Steamships began inspecting boats in accordance with the French Crown's royal ordonnance of January 17, 1846. One of the commission's chief responsibilities was issuing navigation permits for each vessel departing for another port. The permits were granted by a team of mechanics, ranging from two to four men, who oversaw the inspection of the boat's machinery.[46] Boilers, which were prone to explosions, were tested to make sure that they were within the legal limits of pressure allowed, which was four atmospheres.[47] But even the closest inspections could not always prevent tragic accidents.[48]

In spite of the new dangers that it brought, steam navigation boomed during the second half of the nineteenth century. Marseille's own steamer fleet grew from 30 to 201 vessels from 1850 to 1869, as the city took control of 64 percent of France's total steamer tonnage.[49] The number of workers producing steam engines in Marseille rose by five hundred from 1848 to 1866.[50] Meanwhile, along the Mediterranean coast to the east of Marseille, the Provençal naval port of Toulon adapted steam technology to its war fleet. In addition to learning how to navigate in shifting winds, sailors employed by the French military undertook new training regimens to become skilled mechanics and boiler workers, ushering in an age of industrialized naval combat.[51]

An Emerging Sensory Geography of French Regions: Leisure Sailing on the Mediterranean

Among an older generation of sailing captains, the construction of new steamship harbors in Marseille and Toulon during the mid-nineteenth century was met with feelings of sadness. "Steam machines are invading everything," lamented one former mariner, Édouard Paris. "In spite of their known advantages, they nonetheless tend to diminish the genuine qualities of the man of the sea, and every day [the machines] make rarer the opportunities to learn, by practice, the details and the operations of this ancient and poetic navigation by sail."[52] While the sailors of old were versed in the art of tacking to the wind, all that the new mariners could do was to push forward, with no finesse, in a straight line. The result was a labor force that was less talented and more oblivious to their surroundings: "The sail and the rigging, neglected, no longer form the same kind of masters who are so talented in practice as those skilled in mechanical forces."[53] Indeed, by the 1870s, wind-driven coastal cabotage accounted for less than one-fifth of Marseille's commercial activity.[54]

Maintaining a connection to their Provençal sailing heritage thus became, for former mariners, a means of affirming a distinct regional patrimony that was tied to their natural environment. It was no accident that at the very moment when the Mediterranean winds were losing their relevance for maritime commerce in Provence, the editors of *Armana Prouvençau*, the regionalist journal founded by Frédéric Mistral and the Félibrige, chose a wind rose as their frontispiece (see figure 2.9). Designed in 1870 by a sailor, Captain Negrèu from Ceyreste, a small Provençal town not far from Cassis, the wind rose, "La Roso de touti li vènt," hearkened back to an age when winds shaped the rhythms of life in their region. Labeling Provence's prevailing winds in the Provençal language rather than in French, the wind rose was a symbolic act of resistance against the new economic reality that was taking shape. In political terms, the image reminded the viewer of an age when Provence was fully

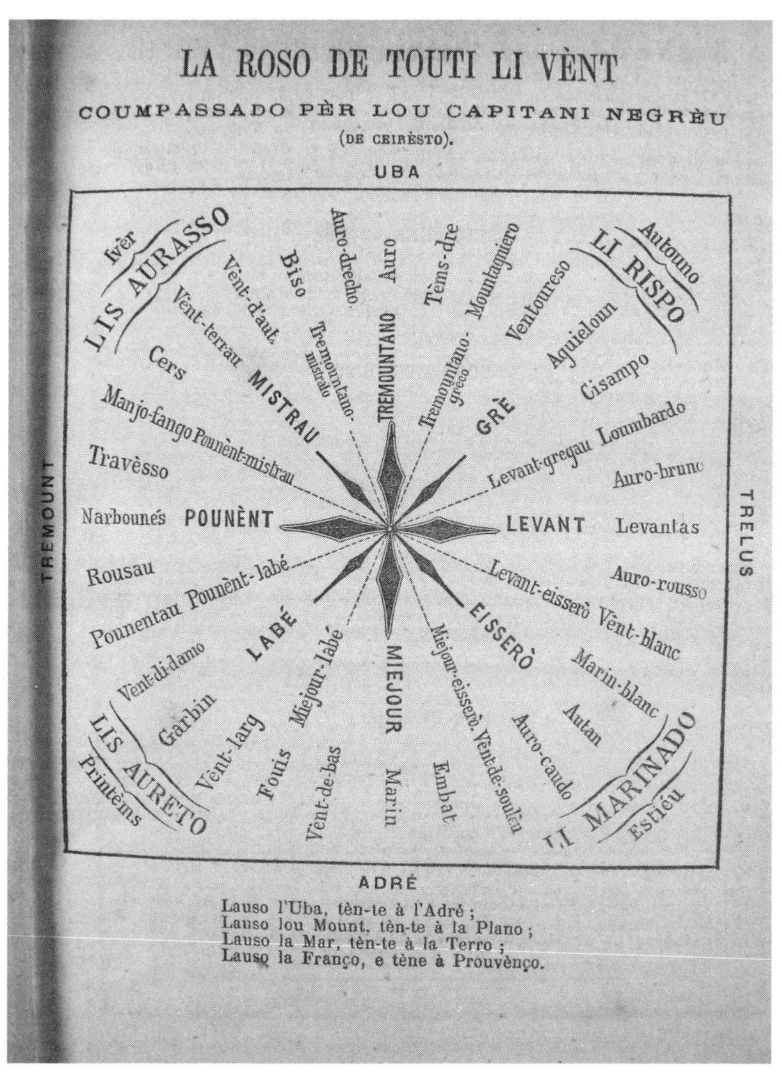

Figure 2.9. "The Rose of All Winds" was the frontispiece image for *Armana Prouvençau*, a Provençal journal that celebrated regional literature, history, art, and poetry. The old-time wind rose gave the journal a nostalgic sensibility, taking the reader back to an age when Provence was firmly situated in a southern European sphere and attached to its Mediterranean winds. "La roso de touti li vènt," *Armana Prouvençau* (Avignon: Enco de Roumanille, 1870). Bibliothèque Méjanes, Aix-en-Provence.

part of a cross-cultural Mediterranean orbit, fused together by the winds that carried sailboats near and far.

It was also in this space of nostalgia for the old regional lifeways of Provence that a new sport rose to prominence: leisure sailing. While the use of sail-powered boats declined steadily among working classes during the late nineteenth century, they simultaneously became objects of status and luxury for middle-class French people. For urban bourgeois men who spent most of their days indoors, shielded from the elements, sailing on the open waters of the Mediterranean Sea offered an opportunity to demonstrate their masculinity through their physical engagement with the wind and the sea. As James C. Williams explains in his essay on sailing as play, "The new recreational focus of sailing provided an escape from the new industrial world . . . Casting off one's mooring lines is an act of liberation from the day-to-day grind and puts one face to face with the forces of nature, with the sublime."[55]

Catering to this new class of sailors were civil associations such as the Nautical Society of Marseille, founded in 1887, which gathered together men of financial means to sail for the "hedonistic pleasure" of it.[56] Sailing guidebooks, aimed at a middle-class audience, offered instructions for how to manipulate the sails and rigging of boats *à voilure latine*, train-

Figure 2.10. Racing yacht leaving the Old Port of Marseille, postcard ca. early twentieth century. © Cd13, Museon Arlaten–musée de Provence.

ing a new generation of French people to recognize and react to the mistral, but this time for the purpose of sport rather than subsistence labor. Some leisure craft were made directly from old fishing boats.[57] Weekend regattas for wealthy Provençaux, as well as for tourists from across Europe, created social spectacles, giving audiences the chance to watch these sportsmen perform their dance with the wind.

Popular visual imagery of the Provençal coastline reflected this new leisure-focused culture of sail (see figure 2.10). Mass-produced postcards from the late nineteenth century often pictured regional scenes that eschewed the reality of steamer boats puffing clouds of dark smoke into the air. Instead, these photographic images commonly centered on a lone sailboat, nestled close to the shore, beckoning tourists to come and experience a return to a simpler, preindustrial age. These portraits of Provence, largely targeting outsiders, framed the region as a place where modern people could find restorative connectivity with the water and the wind. They were images that were recast hundreds of times in the boating scenes found in Impressionist, Postimpressionist, and Fauvist paintings that catered to the lifestyles of the European bourgeoisie.

Conclusion

The invention of steam-powered sea transportation in the nineteenth century resulted in much more than a technological revolution in boat design; it transformed human relationships with nature. French mariners, increasingly reliant on machines, became alienated from the very winds that used to propel them forward. It was a historic turning point, in which scientific intervention "became a tool of deskilling, extracting local knowledge about ecological processes and replacing it with the expertise of scientists."[58] Compared to the *caboteurs* of the past, steamship companies based in Marseille paid less attention to the nuances of seasonal weather cycles and to coastal topography. As the nineteenth century unfolded, the architectural form of ships in Provence followed their emerging function. The increasing efficiency of steam engines led to the removal of the last remaining sail masts from commercial ships as wind power disappeared from all but recreational craft.

And yet, even with the triumph of steam-powered vessels and modern port infrastructure, the mistral continued to exert its might on Provence's coastal communities. A sensational news story from *Le Monde illustré*, published in 1865, recounted a tragic night when the mistral suddenly arrived in wintry Marseille, bringing a frigid blast of air and covering everything with a layer of ice.[59] The mistral's violent gusts descended on the expensive, newly constructed basins to the west

Figure 2.11. Newspaper image of the brig *Le Luxor* and the iron steamer *La Provence* after the two vessels crashed into the Fort Saint-Jean in Marseille during a mistral. The image is based on a sketch by M. A. Crapelet. "Coup de vent du 11 février sur la Méditerranée," *Le Monde illustrée,* February 25, 1865, 117. © CCIAMP / La Collection.

of the city center, where it seized hold of two steam-powered ships—the brig *Le Luxor* and the iron steamer *La Provence*—and tore them from their anchors, sending them smashing against the walls of the port (see figure 2.11). For *La Provence,* it was the steam mechanism in the middle of the craft that hit the quay first, splitting the boat in half before sending it down into the watery depths while its cargo was left to float away in the frothy sea. The following morning, Marseille awoke to a scene of utter chaos and despair. Several lives, in addition to valuable property, were lost.

As the mistral's scene of destruction at the ports of Marseille made clear, weather prediction remained elusive for the modernizing French state well into the nineteenth century. The existing vernacular weather information system in Provence—with its handmade *devino-vènts* and its embodied ways of knowing the weather through outdoor work—was out of step with the demands of expanding national and imperial economic markets that required systematized and reliable weather data. Empowered by its vast bureaucratic reach, the French National Weather Service opened its doors in the 1870s with the goal of understanding the mistral better than the local mariners and farmers who knew its gusts firsthand. It is to this quest for official scientific knowledge about Provence's blustery atmosphere that we now turn.

CHAPTER 3

Ascent into the Wind

THE MONT VENTOUX OBSERVATORY AND THE RISE OF ATMOSPHERIC SCIENCE

Terrifying Ventoux, nest of brown and white eagles,
your bare front, to the south, is white under the snow;
to the north the forest makes a black tail for you . . .
The mistral strikes you and the sun breaks you;
You rise high and proud toward the lightning that tears at you.

THÉODORE AUBANEL, "Le Ventoux"

On a moonlit summer night in 1877, a group of twenty well-to-do gentlemen boarded an omnibus in the sleepy agricultural town of Carpentras and drove nearly two hours to Sainte-Colombe, a village at the foot of Mont Ventoux, the mountain known as the Giant of Provence. There, local guides were waiting with donkeys, which the men used to load up their supplies, and mules, which several of the men saddled to ride. They were now ready to climb. The first segment of the mountain ascent, about three hours long, brought the group to a humble Provençal sheepfold known as a *jas*: a doorless and chimneyless shelter made of simple masonry, with a straw-covered floor. After a short rest, the men got up and continued along the route to the summit at nearly 6,300 feet in elevation.[1] When they finally reached their destination, at four o'clock in the morning, the weather conditions on the mountain peak were rapidly deteriorating. Summer had suddenly become winter.

Reaching for their scientific instruments, the men noted that "the thermometer descended to almost zero degrees" while all around them "the mistral blew with an extreme violence."[2] Attempting to warm themselves, the men gathered in the peak's lone manmade structure, the dilapidated seventeenth-century Chapel of St. Croix, where they lit a fire using the wood that one of the donkeys carried on its back. Undeterred by the wild weather surrounding him, Professor Giraud, secretary of the

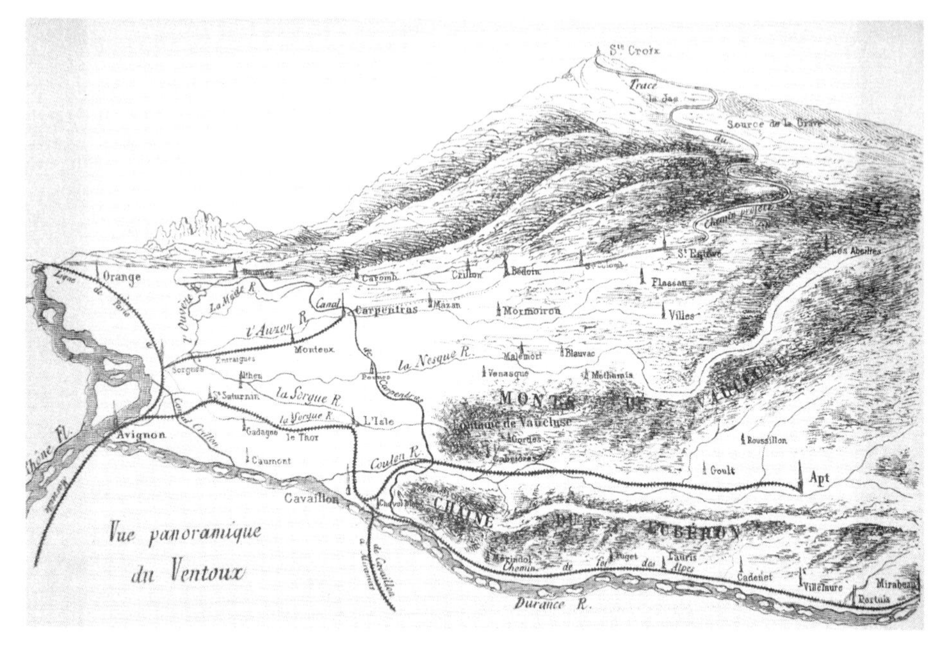

Figure 3.1. Profile view of Mont Ventoux with access routes from the 1870s. "Vue Panoramique du Ventoux," in Bouvier, Giraud, and Pamard, *Le Mont-Ventoux*, 7. Archives départementales de Vaucluse.

Meteorological Commission of the Vaucluse, left the safety of the chapel and began to take atmospheric measurements. Every fifteen minutes, he recorded the direction and force of the wind, the temperature, the air pressure, and the direction of the clouds. "But soon the violence of the wind was such that it was impossible to remain standing on the summit's platform and the instruments were damaged the instant that they were exposed."[3] Though he had successfully climbed the mountain, Giraud had reached "an insurmountable obstacle in the violence of the north wind, and after many tries, he had to give up."[4]

As daylight filtered through the cloud-covered sky, a stunning summit landscape unfolded before the climbers' eyes. Sculpted by the very atmospheric conditions that the men were measuring, the peak of Mont Ventoux—named after the Latin word for "windy," or *ventosus*—was completely bare except for a thin layer of sharp white stones that made walking across its surface "insufferable."[5] There was hardly any vegetation, no animals, and no markers of human existence except for the crumbling chapel. This was a place that right-minded shepherds, hunters, and ice cutters who lived in the villages located in the lower parts of the mountain avoided for much of the year (see figure 3.1). But where savvy mountain inhabitants sensed danger, the meteorologists hailing from the cities below saw an opportunity to discover the secrets of the atmosphere.

Unprotected by tree cover and unimpeded by structures that could slow the movement of air, Mont Ventoux's isolated summit was the perfect location for field research on weather and climatology. "Everyone who took part [in the expedition] was deeply impressed by the extent and the magnificence of the spectacle that they enjoyed at the summit of the mountain," one of the expedition's leaders recalled. They "returned convinced that one could not find, anywhere in France, a situation more favorable for the installation of an observatory."[6] Indeed, just five years after their moonlit summer ascent, the same group of amateur Provençal meteorologists helped lay the cornerstone for the Mont Ventoux Observatory, one of seven high-altitude weather stations built in France during the nineteenth century.[7]

Rising improbably from a barren limestone peak, the observatory at Mont Ventoux, completed in 1884, boldly challenged Provence's mistral-blown atmosphere with its hulking steel-and-concrete structure (see figure 3.2).[8] From the secure base of its impenetrable walls, meteorologists could ascend into the Ventoux windscape and linger in it for days,

Figure 3.2. The observatory on Mont Ventoux was erected on the bare, windswept mountain peak. Alexandre de Bar, "L'Observatoire de mont Ventoux (département de Vaucluse), installé en décembre 1884," in Alfred de Vaulabelle, "Un nouvel observatoire: L'Observatoire du Mont Ventoux," *Magasin pittoresque*, no. 10 (1886): 165. Archives nationales de France.

weeks, and even months at a time. Employing the latest scientific instruments, observers took daily weather measurements and relayed them, via telegraph wires dug into the mountain slopes, to France's central meteorological bureau in Paris. It was there, hundreds of miles away from the Provençal terrain that shaped it, that technicians stripped the mistral of its place-based identity and trapped it within a lifeless two-dimensional paper landscape of statistical tables, graphs, and weather maps. Divorced from its long-standing meaning in Provençal culture by the centralized institutional power of the French weather administration, the mistral lost its very name, becoming simply "the northwest wind" in official scientific terminology.

From the perspective of government officials and scientific experts, meteorology was a foundational pillar in the construction of a prosperous and stable modern French society. The better the nation's knowledge of its weather and climate patterns, they argued, the easier it would be to create a safe and predicable path for advancing key national industries such as agriculture and maritime commerce. However, as this chapter reveals, the French government's plans for establishing systematic national weather observations were thwarted by the unruly behavior of the air itself. Observatories like Mont Ventoux were themselves vulnerable to the extreme physical violence of the wind. Despite their access to sophisticated modern equipment, the staff stationed at the mountaintop observatory found that measuring the mistral's duration, speed, and direction was an ongoing struggle. Destroying the very instruments that were meant to capture it and battering the bodies of observers who were stationed to observe it, the mistral repeatedly eluded the grasp of the French central meteorological bureau. The story the Mont Ventoux Observatory therefore exposes the tensions between a national-scale project to create a uniform system of weather knowledge and a raucous provincial windscape that did not cooperate.

Before the Observatory: Popular Knowledge of the Mistral in Mountain Communities

Long before the construction of the observatory, Mont Ventoux was already home to small communities of ordinary people—hunters, charcoal burners, honey producers, quarry workers, ice cutters, and shepherds—whose economic livelihoods were all impacted by the mistral's presence. Settled in villages tucked safely into the base of the mountain, the inhabitants of rural mountain communities like Bédoin and Malaucène developed seasonal labor strategies and rhythms of movement within the constraints of their physical environment. In order to shelter their bodies during their long walks up and down the mountain, shepherds with herds of sheep—in addition to workers carrying saws, ice picks, and other tools—moved carefully between a network of *jasses*. Related to the French word for shelter, or *gîte*, what the Provençaux called *lou jas* was a "cabin with stone vaults, without doors or chimneys, where anyone passing through could rest on a bed of beech-tree leaves."[9] Before the first road to the summit of Mont Ventoux opened, in 1884, these simple stone-masonry dwellings nestled in the woods offered the only available sanctuaries for travelers who encountered rough weather on their journeys up and down the mountain. In the winter months, the peak, covered with snow, was entirely inaccessible to human visitors.

Their work and mobility closely tied to seasonal weather, communities situated on Mont Ventoux not surprisingly held a special place for the mistral in their sacred cosmologies. The oldest pieces of evidence that we have of human interactions with the mistral in the mountain landscape comes from a cache of clay trumpets, called *li Toutouro*, that French archaeologists uncovered while excavating the summit (see figure 3.3). Ethnographers believe that the musical instruments, which take on a curious twisting form like serpents' bodies, were created by inhabitants of mountain villages during the Middle Ages to magically ward off the unrelenting force of wind and rain with supernatural powers.[10] In addition to the trumpets, evidence of human attempts to coax the mistral into submission can be found in the archaeological remains of an inflatable goatskin ball called *l'Ouire boudenfla*. In the corresponding game, the handcrafted ball was thrown up into the air three times in an attempt to keep the air "prisoner" as long as possible, thereby quelling impending storms.

As Provence Christianized, the use of pagan objects to influence the mistral's behavior overlapped with a new set of devotional practices. For Catholic believers, the swirling atmosphere on the summit of Ventoux became a privileged place to experience the physical presence of God. It was Petrarch, the early Italian Renaissance scholar and poet, who was the first to frame his ascent up Mont Ventoux as a transformative spiritual experience. Upon reaching the summit in 1336, the thirty-three-year-old became overwhelmed by the bodily sensations of nature. "Seized by the unaccustomed intensity of the air and the enormity of the spectacle, I was stunned at first,"[11] he later recounted in letters addressed to Cardinal Giovanni Colonna. Looking down at the Rhône River snaking through the plain below, he reached into his pocket for a copy of the *Confessions of St. Augustine*. A timely gift from the cardinal before his journey, the *Confessions* gave Petrarch a theological framework to express the "holy emotions" that the mountaintop elicited from his soul.[12] It was an

Figure 3.3. Clay trumpets known as *li Toutouro* have been found on the summit of Mont Ventoux. Ethnographers believe that mountain inhabitants used the trumpets' supernatural powers to ward off wind and storms. Later, the trumpets were used at Saint John's Day celebrations, to chase away bad spirits, and during harvest season, to call for the wind's help with shucking grain. This terra-cotta color-glazed example from the Museon Arlaten dates from 1936. © S. Normand - Cd13, Museon Arlaten–musée de Provence

expression of Christian piety that was deeply environmental: an encounter with God that merged the inheritance of theology with the immediacy of a sensory experience of nature in real time.

In the century following Petrarch's ascent, the bishop of nearby Carpentras commissioned a chapel dedicated to St. Croix for the summit of Mont Ventoux. The chapel soon became the destination for an annual pilgrimage on September 14, in which inhabitants from the villages at the base of the mountain would make the challenging four- to five-hour-long procession to the summit. When the weather conditions were difficult, the pilgrims could conveniently turn to the same system of *jasses* that protected shepherds and seasonal workers on their journeys up the mountain. At the summit, the pilgrims would participate in a mass, hoping that the physical strain of their journey and their perseverance in an extreme environment signaled their devotion to God. Once the mass was over, the pilgrims would not stay long. One author noted that the peasants of the region would end their pilgrimage by sliding down the steep part of Mont Ventoux, seated on two boards and deploying a sturdy length of wood to control their speed.[13]

To explain the mistral's genesis, ordinary people in Provence turned to myths, proverbs, and folklore. Likely derived from the tale of Odysseus who kept the wind gods tied up in a mountain in Greece, a familiar myth attributed the mistral's origins to a certain *grotte du vent*, a cave located on the slope of Mont Ventoux from which the mistral supposedly burst forth. So sure were they of the wind's origins that some residents of Provence sent a letter to the prefect, begging him to close the door to the cave to prevent the mistral from blowing.[14] The myth likely contributed to the popular Provençal saying that a prudent person is someone who "knows which hole the wind blows from." Meanwhile, other regional proverbs, based on no factual scientific evidence, promoted the widespread idea that the mistral blows in increments of three, six, or nine days—similar to the cycles of the moon.[15] The closest thing that locals had to meteorological instruments were readily accessible types of vegetation that grew in Provence, such as pine cones and thistles. Depending on whether the vegetation expanded or contracted, people could gauge levels of air flow and humidity. Together, this tapestry of vernacular weather beliefs and practices formed a geographically distinct weather culture that people of the region shared with one another.

The rise of meteorological science during the mid-nineteenth century offered a competing interpretation of the atmosphere. Eschewing the folkloric knowledge of the past, French meteorologists sought to bring Provence out of the "dark ages" of its peculiar regional weather culture and replace it with secular weather knowledge based on universal scientific laws. To do so, they supplanted a seasonally based weather culture with the idea of universal time, requiring standard weather measurements multiple times per day, every day of the year, meaning that observers had to spend the treacherous winter months on the mountain summit, something that the inhabitants of mountain communities found baffling. The need to house the observers, in turn, required a new type of infrastructure far more sophisticated than a *jas*.

"A Great Use for Society": The Rise of Meteorological Science in Nineteenth-Century France

Nineteenth-century French governments saw great benefits to creating a national weather network underpinned by scientific stations like the Mont Ventoux Observatory. In destroying "backward" provincial weather cultures—with their named winds, peculiar proverbs, and magical caves—and replacing them with a coordinated system of observatories working from a shared basis in scientific theory, state officials hoped to create a usable form of weather knowledge that could be applied to a range of national issues. "Sailing and agriculture," predicted one administrator in Paris, "will find [in meteorology] a defense of their interest; but public health, too, is more or less directly dependent on variations in the atmosphere."[16] For "questions touching on agriculture and hygiene, such as the state of waterways, the productivity of the soil, the appearance of leaves and flowers, the cultivation of fruits and trees,"[17] meteorology could provide meaningful answers. Far from a fringe science, meteorology was, in the eyes of the French state, to serve as the cornerstone of a new kind of French society—free from revolutionary disorder but on a sure path of civilizational progress—where citizens would lead increasingly predicable, secure, and prosperous lives.

Investing in meteorological science was not the exclusive idea of the Left or the Right; French governments of different political stripes saw great promise in a centralized weather administration. It was the authoritarian leader Emperor Napoleon III, who had seized power from a democratically elected government through a coup d'état in 1851, who created France's first national weather service. His choice of director for the new meteorological administration, headquartered at the Observatory of Paris, was Urbain Jean-Joseph Le Verrier, a well-known expert in astronomy and climatology. Not unlike the emperor himself, Le Verrier believed in the virtues of strongman rule. An unelected "senator for life," he condemned the "anarchy" that had previously existed in the study of meteorology and called for a rigid top-down ordering of French weather knowledge.[18] One of his first moves, in 1855, was to propose ten observation stations at telegraph offices in different parts of the country, providing each with a uniform set of instructions and instruments for measuring pressure, temperature, and the force and direction of the wind (using a weather vane and observing smoke from chimneys) three times per day. He later expanded these stations to Napoleon III's new nationwide system of universities, the Écoles normales.[19]

The standardized national weather surveys that Le Verrier sent across the country were designed to combat the kind of idiosyncratic, personalized style of weather reporting that had characterized the eighteenth and early nineteenth centuries. As Jan Golinski and Vladimir Jankovic have demonstrated, Enlightenment-era scientific weather observations were carried out by a loosely connected group of cultivated individuals, including provincial doctors and administrators, interested in the relationships among disease, agriculture, and climate.[20] For these well-to-do gentlemen, weather observation was a solo endeavor that satisfied a leisurely intellectual curiosity; the sprawling handwritten documents that they left behind are chock-full of personal anecdotes and local case studies.

In contrast, the method of weather observation that Le Verrier proposed was highly constrained and systematized. His mass-produced weather-observation surveys, with their fill-in-the-blanks columns for data gathering, promoted what Lorraine Daston and Peter Galison have called "mechanical objectivity": a way of doing science that eliminated any mental or physical peculiarities linked to individuals, groups, or humanity in general.[21] The goal of this type of "objective" observation was the creation of "blind sight," a form of "self-abnegation" that denied subjective feelings or impressions of the weather in favor of machine-driven measurements and quantitative reporting.[22]

A number of private citizens active in French civil society welcomed the message of discipline and dispassionate objectivity that characterized Le Verrier's national-scale weather surveys. Like other field sciences that rose to prominence in the nineteenth century—including ecology, biology, and geology—meteorology was widely considered to be "an accessible scientific practice" in which dedicated amateurs could play a vital role.[23] As Fabien Locher has argued, participating in nationwide scientific undertakings allowed middle-class men in particular to demonstrate their personal values of exactitude and meticulousness.[24] For a rising class of provincial people obsessed with signaling their social status to others, owning a set of meteorological instruments was akin to possessing an expensive encyclopedia, atlas, piano, or other objects of bourgeois prestige. The 1850s and 1860s saw a flourishing of scientifically oriented civil associations like the Meteorological Society of France, which aimed to provide a constructive service for the nation while boosting its own members' reputations among their peers.

Partnerships between middle-class meteorological societies and the French state deepened following the revolutionary overthrow of Napoleon III's regime and the aftermath of the disastrous Franco-Prussian War of 1870–71. For members of the French middle classes, the loss to the Germans was a humiliating blow that signaled something broken and deficient about their country—a fear that French civilization was

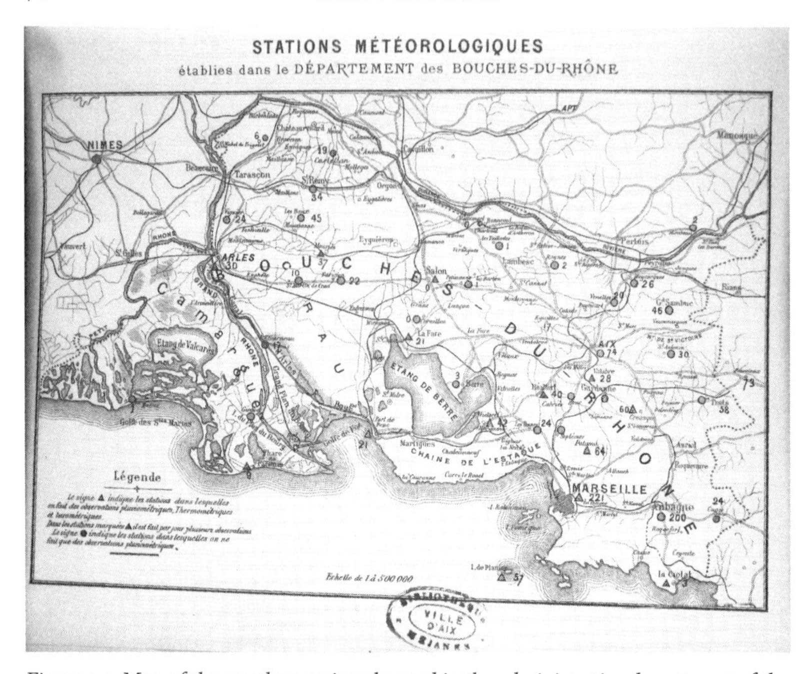

Figure 3.4. Map of the weather stations located in the administrative department of the Bouches-du-Rhône in Provence. "Stations météorologiques établies dans le département des Bouches-du-Rhône," *Bulletin annuel de la Commission de météorologie du département des Bouches-du-Rhône* (Marseille: Typographie et lithographie Barlatier et Barthelet, 1893). Bibliothèque Méjanes, Aix-en-Provence.

lagging dangerously behind its eastern neighbor. France's lack of adequate technological infrastructure became a focal point of their blame. In the wake of the war, the leaders of France's new government, the Third Republic, called for nothing short of the "regeneration of the country through scientific information."[25] Meteorology was one of the branches of science identified as needing major reform. In 1878, the Paris Observatory became its own distinct agency, the Central Meteorological Bureau (*Bureau central météorologique*) under the authority of the minister of public instruction. Besides the reorganization and enhancement of its central weather agency, the nascent republican administration ordered its prefects to establish new citizen-run meteorological commissions for every department in the country (see figure 3.4).

According to an 1879 circular from the minister of public instruction, the well-known republican leader Jules Ferry, each department in France was to establish a volunteer-based meteorological commission tasked with gathering and processing meteorological records for their territory. The commissions were to be "composed of competent and devoted men," who must be selected "among the people whose professions or work would indicate from the start their contribution in an effective manner to the progress of meteorology."[26] As a result, the commissions were overwhelmingly comprised of upwardly mobile members of the new republic, representing a government that rested on the shoulders of a "new social strata" (*nouvelles couches sociales*).

In their published accounts of their high-mountain ascents up Mont Ventoux to conduct weather experiments, Provence's middle-class amateur meteorologists styled themselves as adventurous heroes. Battling and overcoming the power of elements like the mistral was framed as a feat of individual prowess akin to tackling nature in exotic foreign locations. "The climate is harsh at the summit of Ventoux, as harsh as Lapland," Eugène Barrême claimed in his published ascent story.[27] Another climber likened Mont Ventoux to a domesticated African elephant, "who will bend down before those who can mount on his back."[28] In each of these heroic narratives, the amateur meteorologist presents himself as a brave individual who hears "the terrible cries of the winds that fill the ravines" and nonetheless proceeds forward into danger, knowing that "the love of science and the general good are noble goals, capable of conquering obstacles and braving the greatest perils."[29]

Indeed, one of the most attractive features about meteorology for middle-class gentlemen was that it demanded a sort of physical combat with nature that could reveal a man's endurance, character, and will.[30] It was a science that gave republican men a means of proving their honor and manliness through a form of muscular citizenship.[31] Many of the men, for example, emphasized the intensive physical preparation needed before making an ascent. One amateur meteorologist, Félix Achard, recommended that a climber on Mont Ventoux bring "a flannel shirt, a vest, pants made of soft wool, strong boots with spikes, a straw hat, a steel-tipped pole and a flask."[32] He even suggested a specific pace of walking: "One must walk at a slow and extended pace, always in the same state, not with a walking stick, but with a long steel-tipped pole . . . Stop for three minutes every half hour at the beginning, then every fifteen minutes when you arrive at the difficult stages."[33]

In the department of the Vaucluse, the composition of the meteorological commission read like a "who's who" of ascendant provincial men who had taken on leadership roles in middle-class professions. The president of the commission, Henri Bouvier, was the department's chief engineer, while other leading members included doctors, forest inspectors, professors, and museum curators. While their professional lives did not, for the most part, require a working knowledge of the outdoor environment, they nonetheless committed themselves to a civilizing mission of destroying Provence's "backward" weather culture. "There are scientific prejudices, as there are popular prejudices, but in no age have learned men been more disposed to give up theirs, and to submit them, in good faith, to the crucible of experiment and observation," explained Élie Margollé, a leading naturalist and atmospheric scientist from Provence.[34]

Targets of particular ire were farmers' almanacs: the cheap, popular weather guides attached to calendars and sometimes fused with politics and gardening advice.[35] Eugène Barrême, president of the Provence section of the Club Alpin Français and prominent booster of the observatory, called the almanacs the work of "charlatans" who profited from "gullible" readers from the lower classes.[36] The hope was that the Third Republic's newfound culture of secular science would eventually trickle down to lower-class provincials: "The people themselves, too, no longer cling with the same tenacity to their superstitious ideas and demonstrate much greater readiness to listen to the voice of reason," he declared optimistically.[37]

Besides their unflagging commitment to the secular scientific method, what distinguished France's new class of amateur meteorologists from previous generations of gentleman scientists was their awareness of the atmosphere's vast geographic scale. Above all, this new generation wanted to pull meteorology out of "the rut of local observations."[38] In their minds, local weather phenomena like the mistral had to be interpreted as part of much bigger bodies of air. The concept of the atmosphere, Deborah Coen has argued, frequently overlapped with the geography of political power in the nineteenth century; it reflected a widening scale of rulership from regions to nations to global empires.[39] In the words of the prominent nineteenth-century geographer Élisée Reclus, the atmosphere was a vast body "comparable to the ocean, as to the incessant circuit of its waves."[40] To measure the atmosphere's movements, one had to escape the confines of provincial geography. "All meteorologists," explained Henri Bouvier, "are in agreement in recognizing that local observations, distributed on the points placed at low altitudes, are subjected too much to particular influences to be sufficient . . . to deduce the general laws that govern the course of atmospheric phenomena."[41] A high-altitude location was paramount because it allowed for the observation of the grand flows of the atmosphere without local irregularities. As a result, French atmospheric science could escape the limits of provincial weather and expand its reach to much bigger national and even planetary scales.[42]

Fortress in the Sky: The Observatory at Mont Ventoux

As excitement for building a high-altitude observatory spread across Provence, and amateur meteorologists launched a fundraising campaign to help the state pay for it, questions arose about the future building's architectural design. What kind of structure could withstand the power of nature on Mont Ventoux? Together, the French state and middle-class citizens developed a vision for the Mont Ventoux Observatory as a building that could simultaneously protect meteorological observers from extreme weather and bring them into contact with an environment that was "exposed to all of the violence of the atmospheric phenomena."[43] As Henri Bouvier put it: "In order to be able to determine these [atmospheric] laws, it is imperative to operate at high altitudes by placing oneself in a shelter from their influences, in a manner that enables you to directly observe these grand flows."[44] To serve this dual function, the observatory had to take an aggressive stance toward its surrounding environment. In contrast to vernacular forms of Provençal architecture that embodied or mirrored their surrounding regional landscape—the Camargue huts made of wetland reeds or the stone mas farmhouses nestled in the countryside—the observatory planned for Mont Ventoux was a building fit for the modern age: a steel-and-concrete structure modeled on France's handful of existing high-mountain observatories.

Situating the scientific complex within a swirling visual rendering of the rocky mountain summit, the General Plan, published shortly after the building opened its doors, in 1884, reveals the multiple functions that the observatory served (see figure 3.5).[45] In the center of the image, we can see a place for "the former Chapel of St. Croix," which, in a symbolic act of secularization, was demolished to make room for the scientific observatory and reconstructed lower down the mountain. At the bottom of the plan, we see the main observatory building labeled *Bâtiment,* which served as the residence for a permanent staff of observers and a rotating group of researchers carrying out high-altitude experiments. Inside this main building were offices, living quarters, a room for grain storage, and two cisterns with enough capacity to provide water to residents throughout the winter months. The building's exterior functioned as a defensive shield against the mistral. Its north facade was void of any openings and its exterior structure was heavily fortified. "Because of the furious winds that reign up high, it makes sense, in my view, to make a roof in the form of a vault, to make it less vulnerable to the *bise* [mistral] . . . The walls on the exterior must be fitted with a strong coating of limewash and the interior by thin fir branches," suggested Eugène

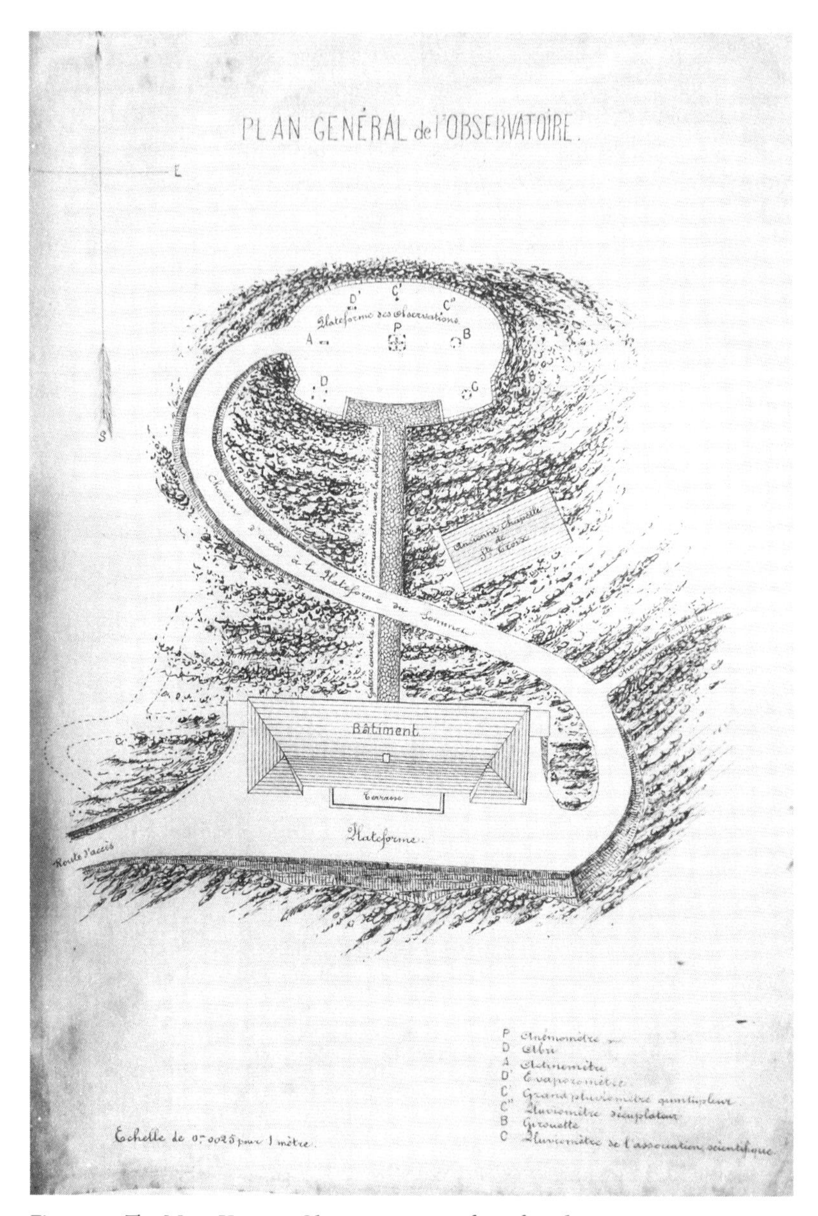

Figure 3.5. The Mont Ventoux Observatory was a feat of modern engineering in an extreme environment. Plan général de l'Observatoire, January 18, 1886. Archives départementales de Vaucluse, 7 M 362.

Barrême.[46] The roof itself was made of a thick layer of hydraulic concrete covered by a double row of cement slabs.

A covered stone tunnel provided access from the main building to the open-air platform where the observatory's scientific instruments, labeled in the plan with alphabetical letters, were located. In contrast to the protective shelter provided by the main building, the platform was entirely exposed to the violence of the mountain weather. The observers' job was to use a variety of mechanical instruments, including barometers (for measuring air pressure), pluviographs (for measuring rain and snow), and anemometers (for measuring wind direction and speed) to track the mountain's shifting atmospheric conditions.

Among weather phenomena, wind is arguably the most challenging to observe and measure.[47] Anemometers, the tools commonly used to record the wind, were notoriously fickle devices that dated back to the fifteenth century. During that time, an anemometer was a rudimentary device that consisted of a hanging disk that swung in the air, providing rough recordings of wind speed. A significant advancement in anemometer technology came in the 1840s, with the invention of the cup anemometer by Irishman John Thomas Romney Robinson.[48] Meteorologists in Provence joined scientists across Europe in adopting the device for their wind measurements. "We evaluated the speed of the wind by means of a cup anemometer that spins at 1,950 turns per minute and corresponds to a pressure of 100 decagrams per 1 kilogram of decimeter squared," read an annual report from the department of the Vaucluse in 1874.[49] But the same report warned of the anemometers' frequent failure when exposed to the elements. "The instruments were frequently broken," the report continued, "and they had to be replaced one after the other."[50]

Despite the well-known fragility of the instrument, Henri Bouvier insisted on installing an outdoor anemometer on Mont Ventoux in 1884 "in a position where the winds would reach an extreme violence."[51] But the scientific tool simply could not withstand the force of the mistral's gusts. The outdoor anemometer installed at the summit platform broke, and thereafter observers had to make wind measurements by hand with a Breguet anemometer.[52] As late as 1898, an inspection report from the central meteorological bureau in Paris revealed that the situation had not improved. "No anemometer," the report stated, "has been able to resist the violence of winds on Ventoux. Observations of speed are made with a hand anemometer that one carries out each time to the higher terrace."[53] The observatory at Mont Ventoux had suffered a failure of technology that was not uncommon in the nineteenth century, a period when the estimated error rate for the measurement of wind speed was 5–13 percent.[54]

Besides its inhospitable environment for outdoor anemometers, the observatory faced another infrastructure challenge: it could not perform its function as a node in France's national weather network until it could communicate with Paris via telegraph. Like the weather instruments, telegraph wires were vulnerable to the material realities of the high-altitude mountain environment. "Unfortunately, in the upper part of the path, the wires, during the winter, were covered with a layer of frost of such a thickness that the telegraph markers were, in large number, turned over by the frequent winds at these heights," explained Alfred Pamard, one of the leading members of the meteorological association. "We had to create an underground line."[55] It was not until a year after the observatory opened, in the summer of 1885, that a telegraph engineer from Marseille was able to get the telegraph wiring to work.

The Mistral Moves to Paris

Once the telegraph system was in place, the virtual distance between the Mont Ventoux Observatory and the central meteorological bureau located in Paris disappeared, as wires carried information in real time across the French countryside. Thanks to the telegraph network, the French capital took on the role of what Bruno Latour has called a "center of calculation": a place where scientific knowledge could be sorted, organized, and distributed in standardized form.[56] "The central bureau," in the words of the Paris Observatory, "exercises its supervisory role and its control over these diverse stations through frequent correspondence and inspections done by a meteorologist designated for the task."[57]

The centralization of French weather knowledge, in turn, demanded a new terminology of weather. While laboring people such as shepherds, farmers, and sailors used Provençal terms and proverbs to describe the mistral, the national weather bureau preferred that meteorologists at the department level adopt a uniform language of wind for all of France. An annual report from the meteorological commission in the department of the Vaucluse, for example, pictures a numbered table indicating the frequency and average intensity of the eight principal winds that blow in the department's geographic area (see figure 3.6).[58] While a person from this part of France would know the northwest wind as the "mistral," in the table it takes on the generic label of NO (*nord-ouest*, or northwest) wind. Below the table stands a figure that represents the various strengths of the different directional winds using simple visual symbols. The arrows in the figure serve as metonyms for the wind, communicating their force through the abstract language of science; the more short, feather-like lines on the end of a staff, the stronger the force of the wind.

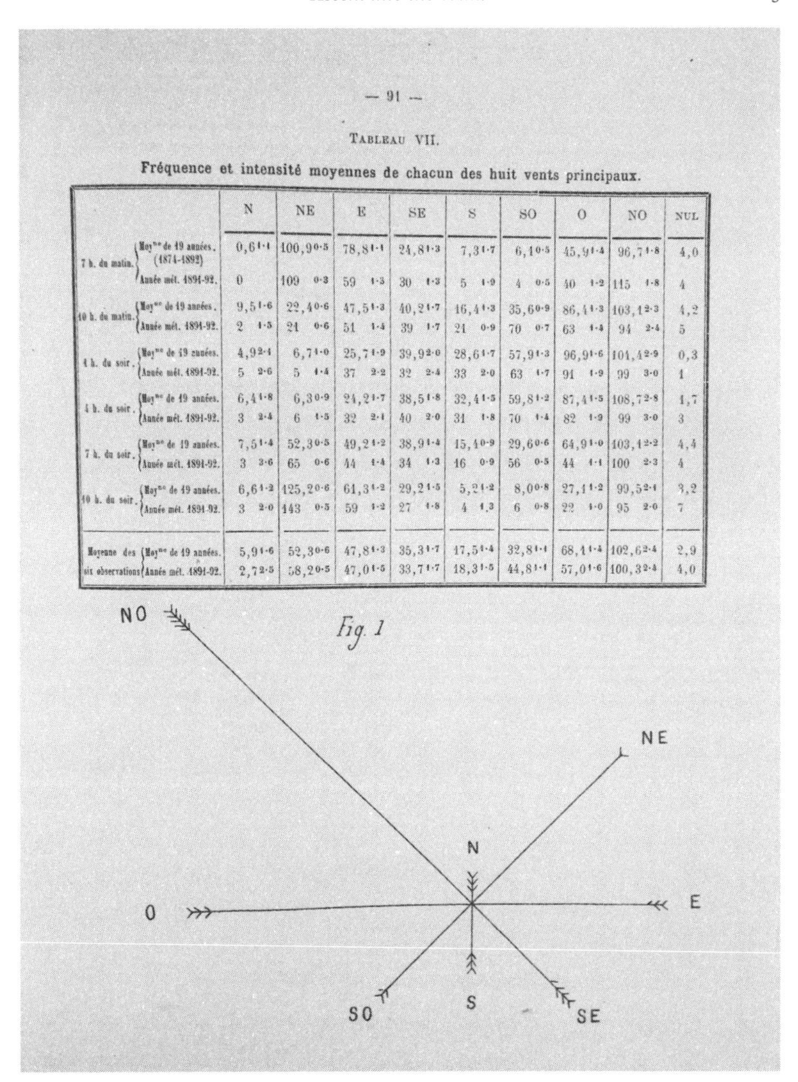

— 91 —

TABLEAU VII.

Fréquence et intensité moyennes de chacun des huit vents principaux.

		N	NE	E	SE	S	SO	O	NO	NUL
7 h. du matin.	Moy^ne de 19 années. (1874-1892)	$0,6^{1\cdot1}$	$100,9^{0\cdot5}$	$78,8^{1\cdot1}$	$24,8^{1\cdot3}$	$7,3^{1\cdot7}$	$6,1^{0\cdot5}$	$45,9^{1\cdot4}$	$96,7^{1\cdot8}$	4,0
	Année mét. 1891-92.	0	109 0·3	59 1·3	30 1·3	5 1·9	4 0·5	40 1·2	115 1·8	4
10 h. du matin.	Moy^ne de 19 années.	$9,5^{1\cdot6}$	$22,4^{0\cdot6}$	$47,5^{1\cdot3}$	$40,2^{1\cdot7}$	$16,4^{1\cdot3}$	$35,6^{0\cdot9}$	$86,4^{1\cdot3}$	$103,1^{2\cdot3}$	4,2
	Année mét. 1891-92.	2 1·5	21 0·6	51 1·4	39 1·7	21 0·9	70 0·7	63 1·4	94 2·4	5
1 h. du soir.	Moy^ne de 19 années.	$4,9^{2\cdot1}$	$6,7^{1\cdot0}$	$25,7^{1\cdot9}$	$39,9^{2\cdot0}$	$28,6^{1\cdot7}$	$57,9^{1\cdot3}$	$96,9^{1\cdot6}$	$101,4^{2\cdot9}$	0,3
	Année mét. 1891-92.	5 2·6	5 1·4	37 2·2	32 2·4	33 2·0	63 1·7	91 1·9	99 3·0	1
4 h. du soir.	Moy^ne de 19 années.	$6,4^{1\cdot8}$	$6,3^{0\cdot9}$	$24,2^{1\cdot7}$	$38,5^{1\cdot8}$	$32,4^{1\cdot5}$	$59,8^{1\cdot2}$	$87,4^{1\cdot5}$	$108,7^{2\cdot8}$	1,7
	Année mét. 1891-92.	3 2·4	6 1·5	32 2·1	40 2·0	31 1·8	70 1·4	82 1·9	99 3·0	3
7 h. du soir.	Moy^ne de 19 années.	$7,5^{1\cdot4}$	$52,3^{0\cdot5}$	$49,2^{1\cdot2}$	$38,9^{1\cdot4}$	$15,4^{0\cdot9}$	$29,6^{0\cdot6}$	$64,9^{1\cdot0}$	$103,1^{2\cdot2}$	4,4
	Année mét. 1891-92.	3 3·6	65 0·6	44 1·4	34 1·3	16 0·9	56 0·5	44 1·1	100 2·3	4
10 h. du soir.	Moy^ne de 19 années.	$6,6^{1\cdot2}$	$125,2^{0\cdot6}$	$61,3^{1\cdot2}$	$29,2^{1\cdot5}$	$5,2^{1\cdot2}$	$8,0^{0\cdot8}$	$27,1^{1\cdot2}$	$99,5^{2\cdot1}$	3,2
	Année mét. 1891-92.	3 2·0	143 0·5	59 1·2	27 1·8	4 1,3	6 0·8	22 1·0	95 2·0	7
Moyenne des six observations	Moy^ne de 19 années.	$5,9^{1\cdot6}$	$52,3^{0\cdot6}$	$47,8^{1\cdot3}$	$35,3^{1\cdot7}$	$17,5^{1\cdot4}$	$32,8^{1\cdot1}$	$68,1^{1\cdot4}$	$102,6^{2\cdot4}$	2,9
	Année mét. 1891-92.	$2,7^{2\cdot5}$	$58,2^{0\cdot5}$	$47,0^{1\cdot5}$	$33,7^{1\cdot7}$	$18,3^{1\cdot5}$	$44,8^{1\cdot1}$	$57,0^{1\cdot6}$	$100,3^{2\cdot4}$	4,0

Figure 3.6. Scientific images offered simple, orderly, and standardized representations of disorderly forces of nature, promoting the idea that weather was knowable and manageable. Fig. 1 in *Bulletin annuel de la commission du département des Bouches-du-Rhône, Année 1892* (Marseille: Typographie et lithographie Barlatier et Barthelet, 1893), 91. Bibliothèque Méjanes, Aix-en-Provence.

The most persuasive images disseminated by the Paris Observatory—and the ultimate symbols of weather's displacement from living provincial environments to two-dimensional pieces of paper—were the national meteorological maps that began to circulate across France in the last two decades of the nineteenth century. Presenting God's-eye

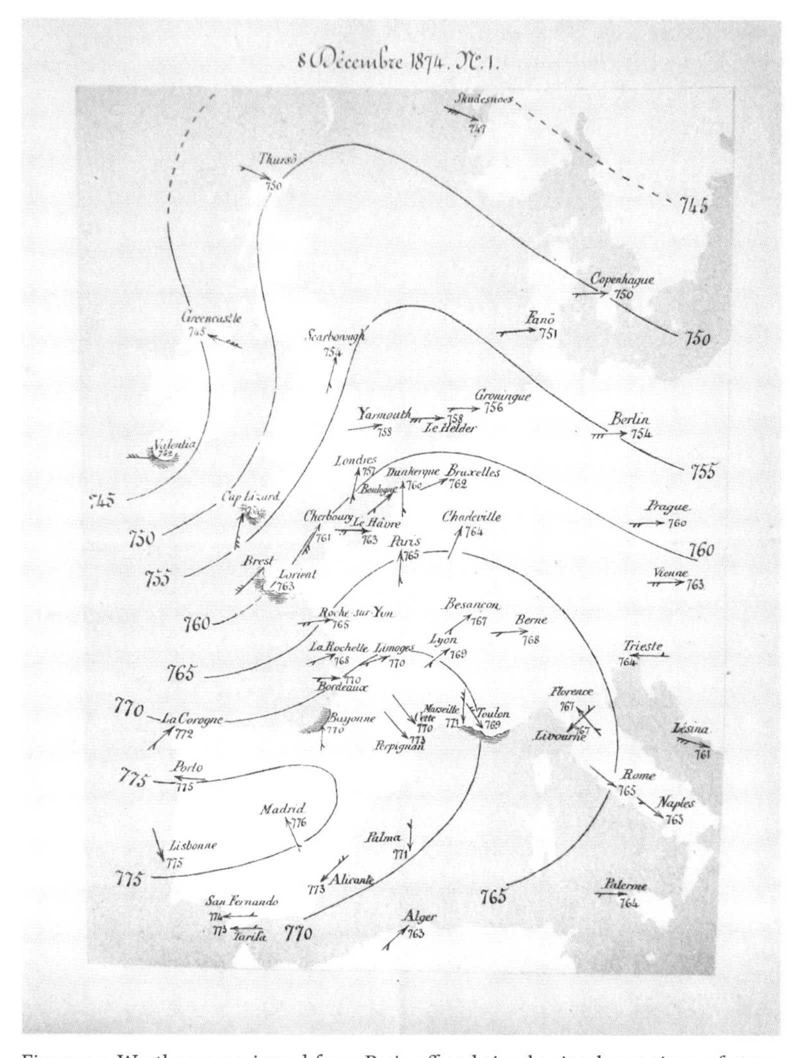

Figure 3.7. Weather maps issued from Paris offered simple visual overviews of atmospheric movements across the entire country. "8 Décembre 1874, No. 1," in Ministère de l'Instruction publique, Observatoire de Paris. VIII. Avertissements météorologiques, 1874. Archives nationales, F17 3727.

views of France marked with simple lines indicating the speed and direction of wind gusts, shifts in pressure zones, and other climate information, the weather maps produced by the central meteorological bureau simply and elegantly transported the atmosphere from the realm of the invisible to the realm of the visible (see figure 3.7).[59] The goal of their visual language was to turn something as dynamic as moving air into a simple, motionless, freeze-framed image. The maps' power lay, as Fabien Locher argues, in their ability to create "an instantaneous vision of the atmosphere."[60] Denying the violent and unruly nature of weather, they promoted restricted, constrained, and disciplined visual impressions of the world that existed above France.[61]

But hidden beneath the nineteenth-century weather maps' illusion of environmental mastery is a more complicated and nuanced story of the scientific infrastructure that existed on the ground at observatories like Mont Ventoux's. Atmospheric research at high elevations was far messier and more dangerous than meteorologists expected or than weather maps let the public believe. Rather than triumphing over provincial nature and revealing its secrets, nineteenth-century meteorologists on Mont Ventoux found themselves confronting a moving force of nature that was difficult to harness and control. The neat and orderly visual products of meteorological science—the weather statistics, charts, and maps—masked a challenging and even grueling embodied experience of scientific labor in windswept outdoor environments.

The Provane Brothers: Scientific Labor amid the Violence of the Mistral

Field science usually takes place in multiuse environments, where ordinary people such as farmers, shepherds, fishers, and tourists coexist with observers in the landscape under study.[62] In the case of Mont Ventoux, a modern scientific complex was built in a place with an existing pastoral community in the village of Bédoin (also Bedouin), which made the decision in 1879 to cede a hectare of communal forest for the establishment of an observatory.[63] From the village mayor's point of view, the loss of territory was worth the modern amenities that the observatory would bring with it: a new road, a telegraph service, and a place to store silkworms in the winter months. It also would bring job opportunities to people in the village. Despite their triumphant narratives of ascending Ventoux and braving the elements in the name of science and the French Republic, the middle-class meteorologists who initiated the observatory's construction had no intention of running the year-round operations there. For this task, they would need to hire permanent observers suitable for spending months at a time on an isolated and windy mountaintop. They would need to hire people from Bédoin.

In considering the ideal type of person to fill the role of a year-round weather observer, the meteorological commission in Provence was influenced by popular beliefs about the correlation between class background and physical endurance to climate. During the long winter

Figure 3.8. Observers who lived at the summit of Mont Ventoux year-round suffered severe physical stress from the extreme weather conditions on the mountaintop. *Mont Ventoux sous la neige*, postcard ca. 1900. Archives départementales de Vaucluse, 21 Fi 33.

months, when the observatory was separated from the rest of the world by snow, the observer would have to rely on the stockpiles of food and water, stored in cisterns, to survive (see figure 3.8). Outside, the icy wind, according to one observer, "felt like pins and needles sticking into you."[64] In the minds of department meteorologists, an ideal candidate to serve as observer was someone who was "acclimated" to such a climate. When Pierre Provane of Bédoin was hired for the post in the mid-1880s, his personnel report stated that he had "all of the desirable aptitudes to fill this post well. He is the son of the head road builder in Bédoin [Gabriel Provane, who oversaw the construction of the new road from Bédoin to the summit] and as a result is used to the climate and is best suited to put up with the lifestyle of an observer because he will be close to his family."[65] Later, when Pierre left the post to complete his military service, his younger brother, Paul Provane, was similarly described in personnel reports as having a body that was suitable for the job: "He is, further, perfectly accustomed to a stay at the observatory, and his robust health permits one to assume that he will continue to withstand the rigorous temperature on Mont Ventoux."[66]

While the ideal weather observer for the mountain observatory was physically sturdy, tough, and acclimated to a harsh climate, he was not overly educated. His task was to step outside to the observation plat-

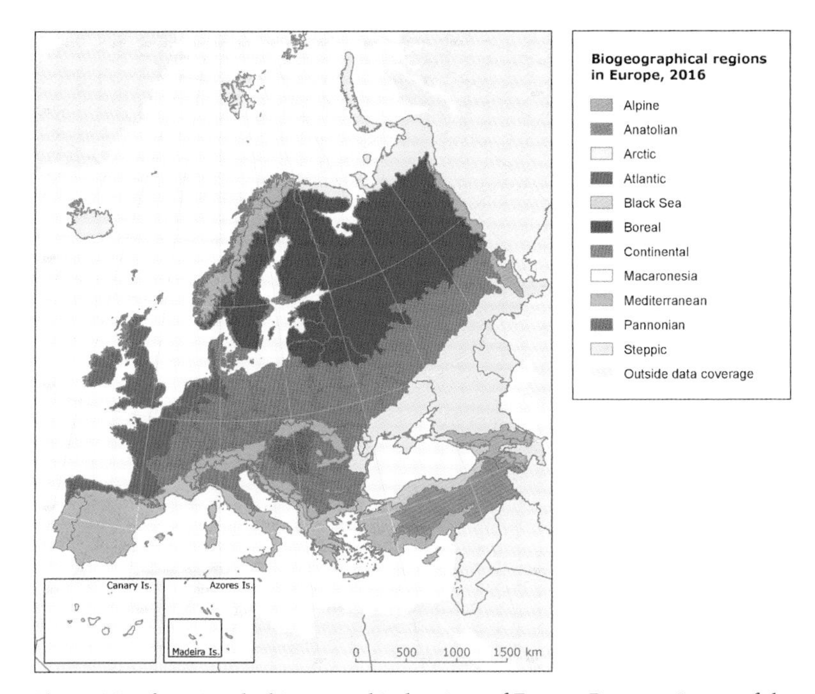

Plate 1. Map featuring the biogeographical regions of Europe. Provence is part of the golden-yellow-colored Mediterranean zone, meaning that its ecosystem is more similar to parts of Spain, Italy, the Balkans, Greece, and Turkey than it is to northern or central France. European Environmental Agency.

Plate 2. Farmhouses in rural Provence were specifically designed for their windswept environment. Walls facing the northwest were always windowless to prevent damage from the mistral. For Paul Cézanne, the farmhouses found on the dry hillsides surrounding Aix-en-Provence embodied the unique character and spirit of his native region. Oil on canvas. Paul Cézanne, *Houses in Provence: The Riaux Valley near L'Estaque*, ca. 1883. Courtesy of the National Gallery of Art, Washington.

Plate 3. One of the most famous boat portraitists of his time, Antoine Roux lived in a building at the Old Port of Marseille, where he sold paintings of sailing craft. His tableaux captured the spirit of sailing in its heyday, shortly before steam power came to dominate Mediterranean travel. This 1801 watercolor, titled *Study of a Sailboat*, reflected Roux's interest in capturing sailing vessels in motion.

Plate 4. Émile Loubon, *View of Marseille from the Aygalades on a Market Day*, 1853. Oil on canvas.

Plate 5. Claude Monet, *Cap d'Antibes, Mistral*, 1888. Oil on canvas. Photograph © 2024. Museum of Fine Arts, Boston.

Plate 6. Paul Gauguin, *Arlésiennes (Mistral)*, 1888. Oil on jute canvas. Art Institute of Chicago.

Plate 7. Vincent van Gogh, *Cypresses with Two Figures*, 1889. Oil on canvas. Collection Kröller-Müller Museum, Otterlo, Netherlands.

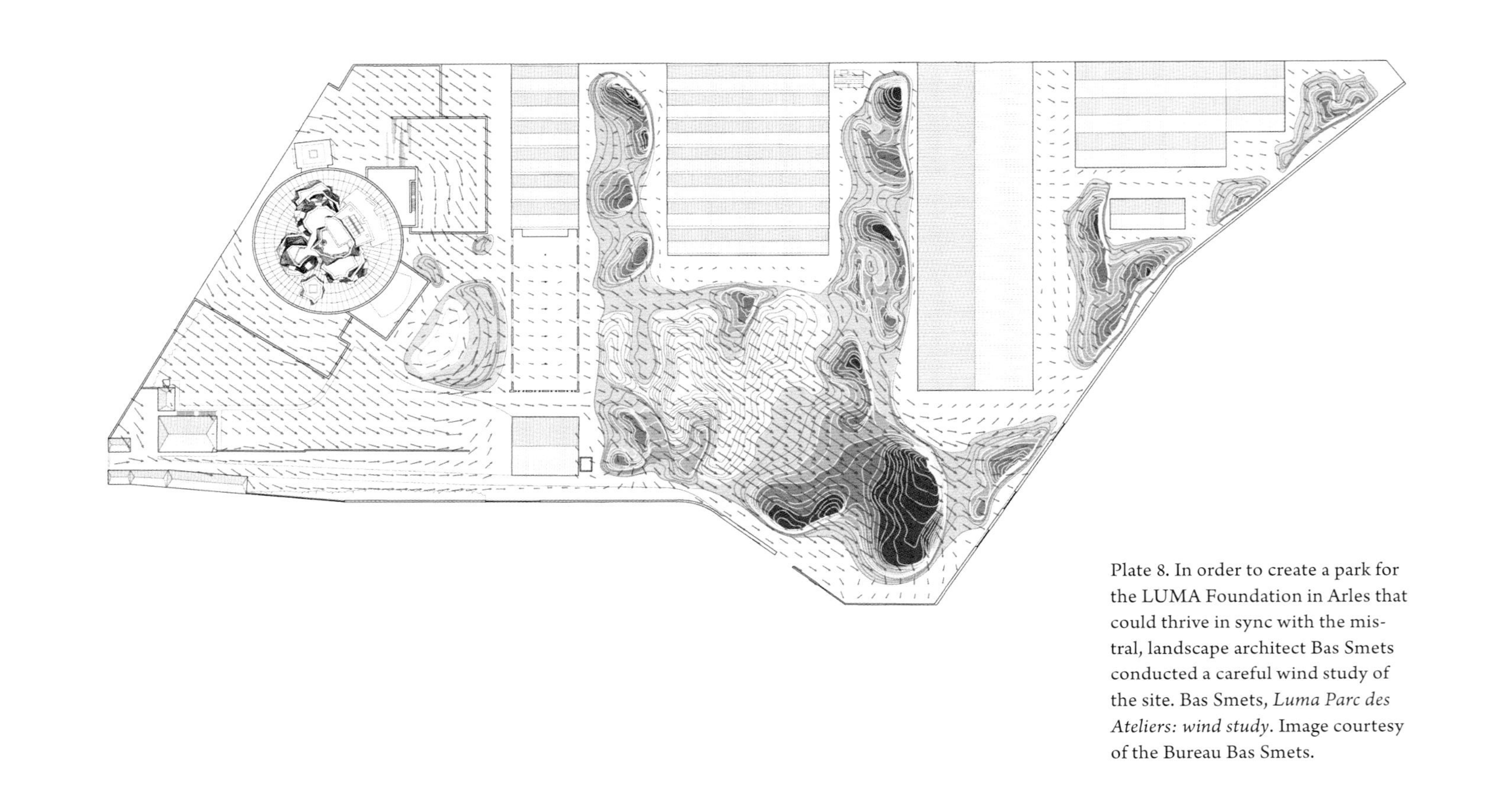

Plate 8. In order to create a park for the LUMA Foundation in Arles that could thrive in sync with the mistral, landscape architect Bas Smets conducted a careful wind study of the site. Bas Smets, *Luma Parc des Ateliers: wind study*. Image courtesy of the Bureau Bas Smets.

form, let the instruments do their work, and report back their readings. It was a form of what Locher calls "industrial scientific work," a form of repetitive manual labor for which the department's meteorological commission determined several generations of the Provane family from Bédoin were ideally suited.[67] As Paul Provane's hiring report states:

> This candidate is only nineteen years old, but he has been attached to the Observatory for two years as an Observer's aid, and he is perfectly skilled with the telegraph and the method for making observations, which do not demand, after all, special technical instruction. What is needed is a great deal of exactitude and care, which, we are convinced, Paul Provane does not lack.[68]

An inspection report from the central meteorological bureau in Paris echoed this assessment, noting that the brothers' lack of scientific knowledge did not impact their ability to make recordings: "The service is entrusted to two brothers Provane, sons of a railway mender in Bédoin, and both teachers . . . Their scientific instruction is a bit cursory, but they carefully make six observations each day with diverse instruments."[69] A mind that could obey orders, rather than a flexible or creative mind, was what the observatory needed.

Indeed, unlike the middle-class republican men from Provence, for whom meteorology was a form of leisure, amusement, and masculine performance, the low-wage observers did not leave behind romanticized or heroic narratives of their experiences in the windswept environment of Mont Ventoux. To get a sense of what their daily lives were like on the mountain summit, we must therefore look to unconventional sources hidden in the archives. Health-care reports from the meteorological office in the Vaucluse offer one key source of information.[70] Many of the reports offer details of the Provane brothers' physical suffering. "The rigor of the temperature together with a constant humidity has caused health problems and I am forced, with great regret, to leave this post,"[71] Pierre Provane wrote to the meteorological commission in the Vaucluse. In 1908, he made a plea for paid medical leave for "a serious illness contracted in the practice of and on occasion of his work tasks."[72] He informed the minister of public instruction that "over the course of his employment, he had contracted, as the attached medical report confirms, a chronic nephritis that renders him incapable of continuing his tasks and prohibits him from working in any other occupation that would allow him to support himself."[73]

In addition to Mont Ventoux's physical toll on the body, the observatory's personnel records document the psychological harm caused by its extreme climate. The question of mental fortitude was a particular concern during World War I, when the departmental weather administration struggled to keep the observatory functioning with the Provane brothers off to war. Local officials decided to install a caretaker who could spend the night there, open the buildings, keep the fire going, and, if possible, take one or two observations per day. A car driver, Xavier Thomas from Bédoin, was offered the job, but would only accept it under one condition. "It is too difficult for M. Thomas to be at the Observatory alone in such isolation," the wartime report stated, concluding, "He must have a companion."[74] Thomas was eventually assigned an Italian agricultural laborer from Bédoin who was deemed suitable for the post despite his non-French nationality. Shortly after the war ended, a doctor determined that Thomas had contracted a "gastro-intestinal illness," attributed to his deficient diet during his isolated winter months at the observatory.[75]

In total, the French state's meteorological bureau relied on an estimated eight thousand to ten thousand *observateurs* to do the work of taking weather observations multiple times per day and telegraphing their measurements to the Paris Observatory.[76] Across Provence, ordinary weather observers like the Provane brothers and Xavier Thomas performed their work quietly. Positions were filled largely by builders and schoolteachers, including women. An observer at the small meteorological observation post located within the village of Bédoin itself, for example, was Mlle Romanowski, who the department's head engineer vetted as "perfectly capable" of replacing her father upon his death, in 1891.[77] In reality, the embodied labor of the state's hired weather observers was not unlike the physical work performed by Provençal sailors or farmers; all had to develop strategies for coexisting with the mistral in order to survive. The major differences in the observers' work environments lay in their clock-based schedules and their placement in fortress-like buildings that trapped them for much of the year. Despite the many sacrifices that they made on behalf of French meteorological science, however, the rural people that helped to run Mont Ventoux were often disdained by their fellow citizens. Alfred Pamard, one of the educated middle-class scientists who helped found the observatory, complained about the rural inhabitants of Bédoin who felt a sense of ownership of the mountain summit and "considered themselves at home [*chez eux*] at the observatory."[78]

Conclusion

Through their nationwide network of observatories, French government officials and scientists created a new system of centralized weather knowledge in the late nineteenth century that effaced the place-based particularity of the mistral, separating the ancient regional wind from Provençal culture and language. Motivated by a desire for progress and technological advancement, this national-scale project was meant to serve the public good—making weather more predictable for agriculture, maritime commerce, and the military—and to prove that France was keeping pace with its European rivals. Eschewing the personalized and embodied knowledge of the weather that had once guided farmers, sailors, and shepherds to safety, this new approach to weather measurement and observation prioritized detached and systematic knowledge over grounded, bottom-up, and regional-scale ways of knowing the wind.

But if we look beyond the records of the state's central meteorological bureau and its departmental administrators, it becomes evident that not all the new research on wind and weather in nineteenth-century France fit within the parameters of a national-scale meteorological project. Experts in the field of French medicine were also hard at work researching the mistral, but from a different vantage point and for a very different purpose. It was precisely the kind of physical suffering that the Provane brothers experienced through a cold winter on Mont Ventoux that drove public health researchers to want to know more about how places and bodily well-being were intertwined in France's distinctive climate zones. Rather than promoting national-scale geographic consolidation, French medical researchers elevated the visibility of regional nature in their studies, suggesting that provincial environments—and their blustery winds—would continue to play a formative role in French lives moving forward.

CHAPTER 4

Good Air, Bad Air

PUBLIC HEALTH AND THE CLEANSING POWER OF THE MISTRAL

> Death and eternal silence would reign over all the earth if it were deprived of the atmosphere that envelops our planet . . . All living beings, whether they walk, climb, or fix their roots in the soil, are not the less children of the atmosphere.
>
> ÉLISÉE RECLUS, 1874

As its cool air descends from the mountainous regions of Provence to the Rhône River valley and the low-lying Camargue wetlands, the mistral has an immediate physiological effect on the people it encounters. "The icy breath of the mistral has a sharp chilliness that is unknown to the inhabitants of the north of France," noted Alexandre Dumas in his travelogue about the South of France. "Instead of penetrating through the skin, it seizes on the marrow of the bones and paralyzes you."[1] In his *Memoirs of a Tourist,* the writer Stendhal complained of "a cold and unbearable wind that penetrates the most firmly closed apartments, and irritates the nerves in a manner that gives mood swings to even the most intrepid."[2] Despite the discomforts caused by its freezing gusts, however, the mistral acquired a reputation in the nineteenth century as a wind that saved lives. In their struggle against deadly epidemics—including malaria, cholera, and typhoid—French doctors and public health officials framed the Provençal wind's legendary force as a remedy for diseases that they believed to be airborne. Calling the mistral the *balayeur du ciel,* or "sky sweeper," medical experts promoted a favorable cultural understanding of the mistral as a natural cleanser that chased pestilence away from the southern French atmosphere.

The notion that the mistral's salubrious effects outweighed its painful and unpleasant qualities aligned with much-older forms of medical wisdom in Provence. "Windy Avignon, irritating with the wind, deadly without the wind," one centuries-old proverb warned.[3] Another popular Provençal proverb equated the mistral's healing powers with that of a physician's touch. "When the mistral enters through the window," the saying went, "the doctor leaves through the door."[4] The fact that educated physicians agreed with the wisdom found in Provençal folklore underscored the enduring power of environmental thinking in French medicine well into the modern period. Epidemic diseases, nineteenth-century French doctors were convinced, stemmed from problematic local ecologies that emitted dangerous particles called miasmas into the air. It was an environmentally determined way of understanding the causes and remedies of disease that situated pathogens in what was around, rather than within, the human body.[5]

Thanks to the prevalence of the miasma theory of disease transmission in the nineteenth century, the mistral became bound up in urgent discussions about how to produce more "good air" for people to breathe while eliminating the "bad air" that made people sick. In sweeping across Provence from northwest to south, bringing a refreshing blast of cool mountain air with it, the mistral appeared as an organic antidote to the hot, humid, and pestilential air that festered in the "Ganges-like" Camargue wetlands as well as in contaminated industrial cities like Marseille.[6] This medicalized narrative of the mistral as a health-giving part of the southern French climate stood in stark contrast to that of the sirocco, a wind that blew into Provence from the south, bringing with it warm and sandy air from the French imperial zone in North Africa. By the end of the nineteenth century, a noticeable dichotomy had formed in the region's medical literature between the European-produced mistral—praised for generating "healthy" and "pure" blue skies—and the sirocco, an African-generated wind that threatened Provence with a dangerously "hazy" and "choleric" atmosphere.

The French public health establishment's belief in the human body's permeability to disease-laden air served as the impetus for new building projects in Provence aimed at "fixing" environments believed to produce miasma. It was due to widespread anxieties about the sickly vapors emanating from soggy marshes that the French government drained large portions of the fragile Camargue wetlands during the nineteenth century. Meanwhile, in Provence's largest city, Marseille, the municipal government, under pressure from public health experts, tore down working-class neighborhoods to create broad avenues that let in gusts of the mistral to sweep away the cholera particles believed to be floating invisibly in the urban atmosphere. The goal of these progress-minded civic improvement projects was to mitigate the regional population's exposure to "bad air"—which felt humid and appeared hazy—while opening pathways for "good air" in the form of the cold, dry, and invigorating mistral.

While scientists later disproved the theory that miasmas cause devastating diseases like malaria and cholera—their actual origins lay in mosquito bites and bacteria-infested water—the fact that atmospheric considerations dominated medical research for much of the nineteenth century meant that doctors left behind some of the most detailed observations of winds in Provence. Indeed, when the central meteorological bureau in Paris undertook an ambitious one-hundred-year review of the French climate in 1899, it soon discovered that many of the nation's best long-term meteorological records were made under the impetus of the Academy of Medicine. "Most only exist in one manuscript copy, some are lost, others sent to libraries where they are ignored," noted the minister of public instruction in a letter to French prefects. "But finding them is the only way to reconstitute the history of meteorology in France for at least a century."[7]

This chapter explores a selection of these cross-cutting medical documents that combine the study of the human body with the study of France's regional and local-scale climates. In their professional practice and in their medical research, physicians promoted a decentralized vision of modern France as a mosaic-like territory with a range of climates that each affected the porous bodies of citizens in distinctive ways. At a time when French provincial citizens found themselves increasingly enveloped by the legal, economic, and cultural structures of the centralizing nation-state, the public health system served as a counterweight. By situating French bodies in small-scale places—where they interacted with local topographies, water, and winds—nineteenth-century medical experts maintained a spotlight on the nation's heterogeneous natural environments and their effects on human health.

The Origins of Environmental Thinking in French Medicine

The notion that there is a fluid relationship between a person's body (their interior ecology) and their surrounding natural environment (their exterior ecology) has a long history in the Mediterranean world, dating back to classical Greece. In his fourth-century BC text, *On Airs, Waters, and Places*, Hippocrates—the inspiration behind the Hippocratic oath—counseled physicians to pay close attention to the natural setting in which their patients lived and worked. Along with atmospheric factors such as temperature, humidity, and rainfall, wind was an essential measuring stick for determining whether an environment was deemed sickly or healthful. Specific kinds of diseases, Hippocrates argued, could only be expected to arise naturally from cities facing winds from certain directions.[8] He was also the first to argue that the north wind, which the Greeks called Boreas, was the healthiest of Greece's prevailing winds, because it removes moisture from the air and makes the sky clear and bright.[9] Though many new medical ideas and theories would flow across the Mediterranean and throughout Europe in the centuries to follow, Hippocrates's environmentally rooted philosophy of medicine had a remarkable staying power. Up through the modern period, most European doctors continued to explain human illnesses based on imbalances in nature that they observed in air, soil, water, animals, and plants.

In France, the Hippocratic approach to medical diagnosis and treatment formed the centerpiece of one of the earliest statewide public health initiatives. Organized by a group of reformers within the struggling French monarchy shortly before the Revolution of 1789, the Royal Society of Medicine (1778–93) was tasked with stemming the spread of epidemic diseases, particularly malaria and typhoid, which had been raging throughout the king's realm. Though it was based in Paris and authorized by royal patent, the Royal Society promoted a decentralized approach to medical research that encouraged provincial doctors to explore the links between their local weather conditions and the timing of deadly epidemics.[10]

The result was a series of "medical topographies" that visualized the French realm through the lens of human health. Local winds featured prominently in these small-scale studies. One medical topography of Marseille, published in 1780, dedicated no less than ten full pages to an analysis of the city's windscape.[11] Of particular interest to the study's author, Dr. François Raymond, were two winds, one from the northwest and one from the southeast, which he described as the two dominant winds that alternately reigned over Marseille. Not only did these two dominant winds have particular effects on the atmosphere—one wind was always cold and dry, while the other was humid and brought clouds—but the winds, Dr. Raymond argued, also produced different kinds of physiological reactions in the human body.

Basing his medical topography directly on Hippocrates' ancient texts, Dr. Raymond reported that "in Greece, in Italy, and in Provence, the winds of the north revive the organs, fortify the body, and sharpen the senses and the mind, and the winds of the south have a contrary effect."[12] Praising the overall climate of Marseille, he noted with local pride that "the climate of Marseille, distinguished by its purity, its mild temperature, and principally by its dryness, is a model of comparison with other large cities in Europe." Hippocrates's looming influence was hard to miss in the author's presentation of a wind rose that labeled the city's prevailing winds in the language of the ancient Greeks as well as in the languages of modern Europeans (see figure 4.1).

90 MÉMOIRES DE LA SOCIÉTÉ ROYALE

CARTE DES VENTS,

Tracée d'après les indications d'Ariſtote & de Pline.

N. B. *On y trouve les noms Italiens, Provençaux, Latins & Grecs: les noms Provençaux ſont omis, quand ils ſont les mêmes que les Italiens.*

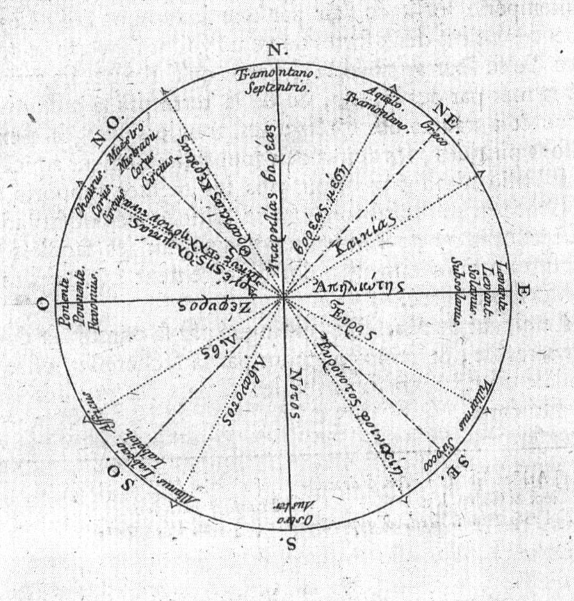

Figure 4.1. Following the Hippocratic method, medical researchers in eighteenth-century France believed in a correlation between disease outbreaks and local weather conditions. François Raymond, "Mémoire sur la topographie médicale de Marseille et son territoire; et sur celle des lieux voisins de cette ville," Société royale de médecine, *Histoire de la Société royale de médecine* (Paris: Imprimerie de Philippe-Denys Pierres, 1780), 90. Source: gallica.bnf.fr/BnF.

While Dr. Raymond touted Marseille's dry and healthy atmosphere, concurrent medical reports from Arles, located near the coastal Camargue wetlands, emphasized the dire state of the latter city's atmospheric conditions. For doctors trained in the Hippocratic approach to medicine, waterlogged and marshy places like the Camargue were the worst type of environments. In his handwritten description of "the state of illnesses in Arles in the different seasons of 1783," Dr. Louis Bret, a member of the city's academy of medicine, claimed that winds had "dispersed the vapor that rises up from our marshes," spreading dirty air to the vulnerable populations living nearby.[13] The main symptom of this illness was "intermittent fevers," which often meant, in the language of the time, an outbreak of malaria.[14]

Dr. Bret's contention that marshlands contributed to malaria outbreaks was not, of course, entirely incorrect. Malaria is spread to humans through bites from mosquitos infected with parasites, and mosquitos do thrive in the Camargue. But instead of looking for malaria vectors in wetland insects, eighteenth-century French doctors like Dr. Bret focused instead on tiny miasmatic particles that they believed to be floating in the air itself. While these particles were invisible, evidence of their existence could be detected in the visual appearance of cloudy, foggy, or hazy skies. To learn more, Dr. Bret became a close observer of the state of the atmosphere; his medical journals included carefully hand-drawn tables indicating the number of rainy, snowy, cloudy, and windy days per year, followed by his hand-scrawled observations about everything from the migrating birds that he saw passing through his area to the date and quality of the local harvest, to the appearance of "intermittent fevers" among townspeople (see figure 4.2). It was a holistic approach to observing Provençal nature and human health in a manner that linked people, disease, vegetation, and the local climate together into an integrated whole.[15]

Dr. Bret's search for the environmental roots of disease in Arles resonated in the writings of French natural historians from the same time. In his extensively researched *Natural History of Provence*, published in 1782, Michel Darluc compared the pattern of disease associated with normal seasonal change and those diseases associated with miasma, a type of air that he considered atypical and out of balance.[16] Addressing the "nature of the miasma," Darluc described "emanations coming from stagnant water in the marshes, which have as much aerial fluids, distinct from atmospheric air, as subtle gases full of volatile acids [that are] toxic and very capable of bringing about the destruction of our organs and even asphyxiation and death."[17] Natural history and disease etiology

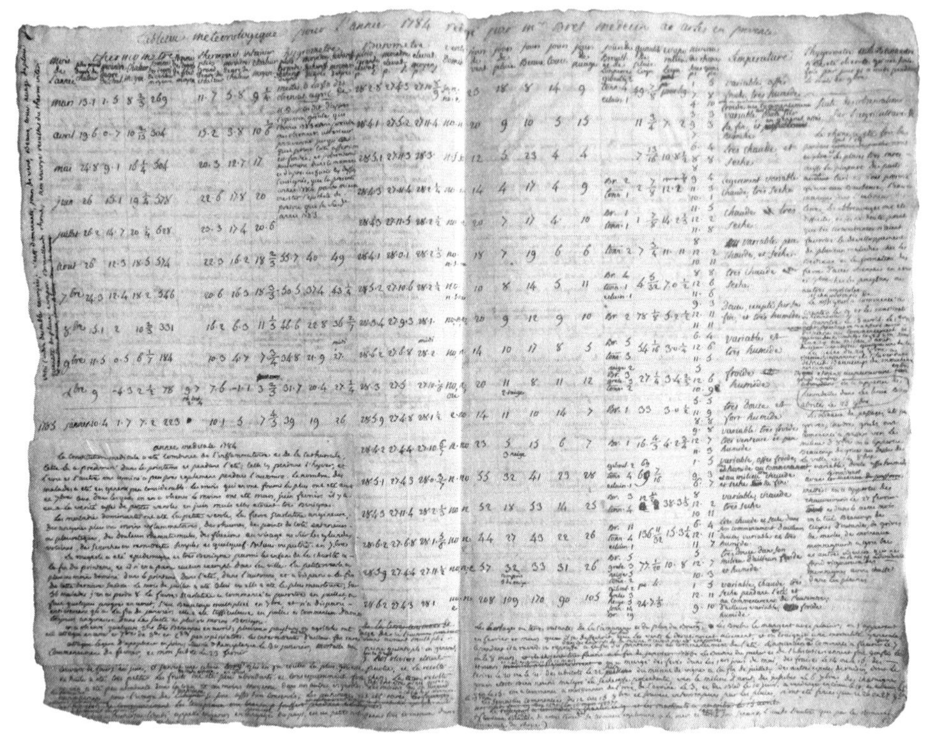

Figure 4.2. Doctors in late eighteenth-century Provence were ardent observers of the weather. Some, like Dr. Bret from Arles, left behind elaborate handwritten charts containing measurements and descriptions of local meteorological conditions in an effort to explain the correlation between atmospheric changes and the prevalence of disease. "Tableau meteorologique pour l'année 1784 rédigé par M. Bret medecin de Arles en Provence." Archives départementales des Bouches-du-Rhône, 1 J 980.

were thus closely aligned in late eighteenth-century France, and both demanded close attention to specific bioregional climates and ecologies.

While the French Revolution of 1789 upended many of the nascent public health initiatives from the ancien régime, including the Royal Society of Medicine, which was disbanded during the Great Terror of 1793–94, the fundamental approaches underlying French medicine—its reliance on place-based Hippocratic ideas—remained unchanged and even flourished. The Revolutionary calendar, for example, implemented in 1793 and designed as a radical reordering of traditional Christian time, embraced the same cycles of seasonal weather that structured French medical research. Many of the months in the new calendar were named after specific weather phenomena—"mist" (*Brumaire*), "frost" (*Fri-*

maire), "rainy" (*Pluvoise*), and "windy" (*Ventôse*)—while other months took their names from the life cycles of plants, including "flower" (*Floréal*), "harvest" (*Messidor*), and "fruit" (*Fructidor*). It was a semi-pagan view of time popularized through revolutionary print culture, including monthly calendars featuring beautiful women in different environments and weather conditions. In their depiction of the "windy" month of *Ventôse*, for example, the artist Louis Lafitte and the engraver Salvatore Tresca pictured a classical-looking female figure in a windblown coastal setting similar to Provence (see figure 4.3).[18] Her livelihood tied to the sea, she carries a fishing rod and a basketful of fish and eels as her hair and drapery flutter in the air.

The seasonally based Revolutionary calendar ordered time in a manner that synched well with doctors' accounts of epidemic disease in Provence throughout the Revolutionary and Napoleonic periods. Take, for example, the handwritten medical observations of Dr. Joseph Guérin of Avignon, a physician and natural historian who wrote one of the first medical topographies of the Vaucluse in northern Provence. His discussion of local winds in his "Meteorological Observations of Year XIII [1804]" followed a progression of seasons. "The great winds of the north and the frosts arose over the course of Germinal, producing strong fears in the farmer," he explained. "The mulberry trees and the stone fruits suffered a great deal, though less in our land than in the neighboring communities that were less protected from the wind." The month of Floréal, he continued, "is again the victim of strong gusts of wind. The weather is cold, the season has hardly advanced, and the first swallows have yet to appear as of the 20th of the month. The leaves of the mulberry appear sad and yellowed. Battered by the wind, exhausted by the cold, they fall, seeing their existence as futile."[19] At the same time, the blustery northern wind, he observed, had a naturally beneficial effect on human health in the Vaucluse. In contrast to the leaves that appeared "battered" and "exhausted" by the cold wind, he asserted that "the air we breathe in this city [Avignon] is generally crisp [*vif*] and healthy [*sain*]."[20]

Like Dr. Bret from Arles, who noted the correlation between malarial fevers and the humid air found in the Camargue wetlands, Dr. Guérin saw dangers in any part of the region surrounding Avignon that he considered excessively wet. "The earth," he wrote in his medical observations from 1807, "was irritated by a great quantity of water that fell in autumn." It was only thanks to "the energy of the sun and the winds" that the earth "cleared the excess humidity that we should have had."[21] In an effort to create the drier environments that doctors

Figure 4.3. The revolutionary calendar reframed time according to annual cycles of weather and plant growth. The period between February 19 and March 20 became *Ventôse,* or the windy month. Salvatore Tresca and Louis Lafitte, *Ventôse* (Paris: Chez l'auteur, 1797). Source: gallica.bnf.fr/BnF.

wanted, the government of Napoleon Bonaparte decided, in the same year, to assert greater control over France's marshy ecosystems. The law of September 16, 1807, gave government authorities permission to dry out areas of the Camargue in a manner "that it judged to be useful or necessary."[22] While plans to canalize and drain large sections of the

Camargue were in part motivated by fears of flooding, Napoleonic officials were also convinced that draining the wetlands would bring about medical progress through the creation of a cleaner, less pestilential, atmosphere.

The Mistral and the Sirocco: A North–South Medical Topography of Winds in the Camargue

With few natural or built barriers, the Camargue is brutally exposed—perhaps more than any other location in Provence—to winds from all directions. When the mistral arrives, it significantly lowers the average temperature of the air in the wetlands in addition to increasing evaporation, decreasing atmospheric humidity, and decreasing the appearance of haze in the sky.[23] The winds of the south and southeast, on the other hand, which constitute 20–25 percent of total wind in the Camargue, are charged with humidity; they increase the haziness of the sky and the amount of moisture in the coastal air.[24] For the civil engineers dispatched to Provence to improve the region's environmental conditions in the early nineteenth century, knowing how these different winds shaped the region's geography was key to determining which future environmental modifications might be possible.[25]

In 1817, François Poulle, a civil engineer working for the French state's Ponts et Chaussés (Roads and bridges) administration, published an environmental study that carefully assessed the Camargue's climate features. "Upon the arrival of the winter solstice," he noted in his *Study on the Camargue*, "there appear, ordinarily, the fiery winds of the north that blow every year at least seven or eight months. They begin, so to speak, their reign with the year, growling and calming themselves successively with each new season . . . Among them, the mistral, or the wind of the northwest, is the most remarkable . . . Its gusts almost never moderate . . . It can acquire a violence capable not only of stripping trees of their fruit but also of breaking their branches, of knocking down roofs and chimneys of habitations, and of generating furious tempests in the water." But the civil engineer was also careful to balance his warnings about the mistral's destructive qualities with a discussion of its beneficial effects on the wetland atmosphere. "The mistral," he explained, "makes up for these grave inconveniences by the same number of advantages. It puts an end to long rains; it dissipates the fogs and the clouds; it dries out submerged fields; it replaces with a healthy air the air that is ceaselessly defiled by the deadly emanations of the marshes."[26]

Not only did Poulle frame the mistral as an advantageous force of nature that benefited people in the Camargue; he also placed the northern wind in opposition to a southern wind that entered the wetlands from the sea. Over the course of the year, he wrote, these two mighty directional winds engaged in an invisible battle for dominance of the skies above the Camargue. "Springtime sees the winds of the north fade and the arrival of the winds of the south. These winds, from the south and the east, pass over the Mediterranean, chasing the humid vapors that, stopped in the north by the mountains, spread out progressively across the plain." These southern winds engage in "a relatively long struggle with the winds of the north that they have replaced [and] the winds of the southeast disappear until the autumn equinox." Their "feeble suffocating breath, pushing deleterious vapors that rise up from the marshes toward the Continent, produces in men anxiety and depression. . . . We owe to [the southern winds] unhealthy humidity in the nights and thick fogs from which the Camargue too often suffers." The struggle between northern and southern winds continued into the winter months until the mistral reasserted its dominance and, as a result, "The blue of the skies radiates in all of its purity."[27]

Completed just a couple of years after Napoleon Bonaparte's fall from power, Poulle's study on the Camargue was written in the context of an emerging French imperial presence in Africa, including the recent invasion of Egypt. The southern wind that Poulle referenced was commonly known in Provence as the "sirocco" (also "syrocco" or "scirocco"), which first appeared in climatic and medical descriptions of the Mediterranean region in the 1600s.[28] The term entered the French language from Italian, and Italian sailors likely borrowed the term from the Arabic *sharqi*, meaning "eastern."[29] A warm southeasterly wind that usually sweeps upward from the hot and dry Sahara or Arabian deserts, the sirocco typically arrives on northern Mediterranean shores in the spring. Due to its desert origins, the sirocco is frequently dusty, and it often brings humid and hazy air thanks to the moisture that it picks up as it crosses the Mediterranean.[30] According to the widespread cultural perceptions of air at the time, the sirocco therefore had the unmistakable look and feel of miasma.

As historian Michael Osborne has argued, the miasma theory of disease transmission in nineteenth-century France grew beyond classical Hippocratic models to include medical observations from non-European places across the globe.[31] The more that French naval hygienists and other imperial travelers studied tropical climates and compared them with more temperate European zones, the more convinced they became of the dangers of hot air. The stereotype of African air as bad for human health helped European nation-states to tout the naturally salubrious nature of their own atmospheres and to extol the virtues of dry and cool winds like the mistral. In France as in Great Britain, "It seemed necessary that infectious diseases had to be imported from some sultry climate."[32]

The perceived differences in the physiological effects of European-generated northern winds and African-originated southern winds in Provence would grow even stronger in the coming decades of the nineteenth century, when cholera epidemics swept across the region. Compared to the modest number of deaths recorded in doctors' journals from the Revolutionary and Napoleonic eras, the cholera epidemics that claimed thousands of lives in the region from the 1830s onward were far more devastating. Because cholera struck cities with a particular vengeance, medical topographies in Provence shifted their geographic focus from sparsely populated wetland areas, like the Camargue, to urban centers. The scope of the studies, however, remained small, focusing on the day-to-day interactions between people in Provence and the volatile zone of air that surrounded them.

Infected Air: The Cholera Miasma in Marseille

Endemic to India, cholera is a fast-moving disease that causes severe vomiting and diarrhea, leading to a loss of up to 25 percent of a victim's body fluids. As a result, a victim's "skin became blue and corrugated, the eyes sunken and dull, hands and feet cold as ice."[33] Most victims died within three or four days. Cholera first arrived in Marseille in 1834–35, killing 3,441 people at a time when the city had grown to become Europe's third-largest port and a bustling hub for Mediterranean trade, particularly after the French army's conquest of Algeria just a few years earlier.[34] While city officials had already built a health management office (*intendance sanitaire*) at the Fort Saint-Jean for the express purpose of monitoring the comings and goings of ship passengers in order to prevent the spread of plague, the cholera epidemic invaded the city with a force that seemed impossible to stop. In a race against time to diagnose and treat the illness as quickly as possible, local Provençal physicians—all trained in an environmentally driven, Hippocratic approach to disease—immediately searched for something amiss in the air.

In his postmortem account of Marseille's 1834–35 cholera epidemic, physician Louis Méry focused on unusual occurrences in the sky. "The city," he noted, "was little by little invaded by fetid vapors; fogs that were thick and low filled the atmosphere; they came from the coast of Toulon and they seemed to follow the coast. [The vapor] was so thick that the Tower of Saint Jean disappeared in a mass of fog."[35] Before long, "The sky lost its elasticity, it weighed heavily on our heads, the vapors of the evening and the morning seemed menacing, and already the people with nervous temperaments submitted to a terrible malaise."[36] But death, according to Méry, did not arrive until the winds began, stirring the poisonous atmosphere and spreading cholera across urban districts. In his account, he identified hazards in both the northern and the southern winds. Both, he understood, had channeled cholera particles around the city, though he considered the latter to be more deadly. "We will note," he argued, "during the period of the first cholera, that the outbreaks of the epidemic in Marseille were always manifested by a northwest wind (the mistral) . . . During the second period, for the days that were more deadly, it was the winds of the southwest that dominated."[37] In an attempt to reduce citizens' bodily exposure to the poisonous winds, the municipal government suspended all commercial activity. But it was too late. "[Cholera] was in the air, in the clouds that, in the night, thundered in storms . . . The wind, in passing by this mass of dead people, dispersed, in the air, the infected odor of the tombs."[38]

As Dr. Méry's analysis underscores, French medical experts from the 1830s saw cholera as a disease that originated outside the human body, in its surrounding environment.[39] Like the eighteenth-century doctors who associated malarial fevers with the Camargue wetlands in Provence, physicians linked the cholera outbreak to Marseille's filthy urban landscape, which emanated putrid air that smelled bad and seemed ripe for carrying disease. As Dr. Méry put it, the cholera outbreak was a two-way street in which both natural and manmade ecologies were responsible. "The [cholera] invasion," he stated, "was the effect of air on the masses of men, and [the effect of] the masses of men on the air."[40] Airs, according to this "miasmatist" (or contagionist) school of medical thinking, could be organized in a hierarchy of good to bad: "[The air] that you breathe in a hospital is less healthy than the air of a city, the air of a city is less healthy than the air of countryside."[41] Yet even as French doctors in the 1830s were convinced that disease particles entered into the human body through breathing unclean air, the precise material form of the cholera miasma—its physical qualities as a substance—remained mysterious. While they were convinced that the cause of cholera was "a deadly agent suspended in the atmosphere," the doctors could not manage to positively identify it.[42] Nor could they establish precisely how, once absorbed into the air from infected soil or water vapor, the deadly particles migrated from place to place.

In order to make the correlation between the spread of disease and changes in local air quality, the French government turned to the power of its growing national bureaucracy and its increasingly sophisticated methods of statistical analysis. The Second Republic's decree of December 18, 1848, required each department in France to establish a council of public hygiene and health.[43] Then, shortly after his coup d'état in 1851, Emperor Napoleon III used this newly established departmental health infrastructure to distribute epidemic circulars to prefects across France, ordering them to facilitate place-based medical studies. The circulars asked doctors in plague-stricken communities to fill out information about local topography (quality of the soil, existence of woods, existence of waterways and their quality) and the local climate, including information about "winds that regularly prevail; that prevailed during the epidemic."[44]

Knowledge gathered from the central government's localized epidemic studies was put to the test in 1865, when Provence was struck by another major cholera outbreak, which claimed three thousand victims in Marseille alone.[45] When the disease first hit the city, in early summer, public opinion initially blamed a group of Muslim pilgrims for the "crime of importation of cholera."[46] In mid-June, a ship named *The Stella* had dropped anchor at the Port of Marseille, arriving from the Port of Alexandria, carrying a number of Arabs who had recently traveled to mosques in Mecca and Medina. Despite a media storm, medical experts from Marseille fought back against the rumor that Muslim pilgrims had brought the plague into the city, arguing that it was surely *atmospheric problems* that had caused the outbreak instead. "The genesis of cholera," argued Dr. Pierre-Auguste Didiot, one of the city's leading medical authorities, "must reside in the element the most universally widespread, the atmosphere, [which is] the only thing that can give us the reason not only for [the epidemic's] generation and its generalization, but also for its prompt diffusion."[47]

Far more significant than the arrival of Muslim pilgrims from Alexandria, argued Dr. Didiot, was the fact that the epidemic had been proceeded by an abnormal, "choleric" atmosphere in Marseille. The weather in the spring of 1865, he noted, had been unusual for its high temperatures, absence of rain, infrequent winds, humid nights, and low barometric pressure. Over time, this intemperate atmosphere slowly acquired a "choleric constitution" that resulted in "an extraordinary calm in the winds, of the more or less pronounced deviation from their usual direction, and an exceptional serenity of the sky."[48] Deadly detritus was thus allowed to accumulate in the urban air, ready to asphyxiate the population. It was a kind of fetid and stagnant atmosphere, according to Dr. Didiot, that was distressingly atypical for the South of France and was far more similar to the air found in a country "from the tropical zone."[49] Influenced by a growing body of colonial medicine, he noted that "the miasmas of Marseille, Toulon, Arles, and Paris could acquire like those of the Ganges a choleric property under the general influence of an atmospheric constitution."[50]

Though rare, a small number of French scholars pushed back against the binary of "good" and "bad" winds that dominated medical discourses on the atmosphere in the nineteenth century. The geographer Élisée Reclus, for example, an anarchist who rejected the world's division into bounded nation-states, preferred to interpret winds as vehicles of interconnection and exchange among global cultures. It was thanks to the dust brought by the sirocco that, according to Reclus, "The northern coasts of the Mediterranean are brought nearer to the great deserts of Africa."[51] Rather than framing the mistral and the sirocco as separate directional forces engaged in combat, he saw Mediterranean winds as working together to achieve atmospheric balance. "Those parts of the world united by atmospheric currents become thereby neighbors to each other," he insisted. "By the incessant mixture of the aerial masses all regions of the solid kernel of the earth are brought nearer, contrasts blended, and harmony is established."[52]

But the dominant views of "good air" and "bad air" promoted by the French medical establishment held tremendous sway, contributing to the wholesale transformation of French cities and towns in the mid to late nineteenth century. As French urban planners brought a spirit of rational design, efficiency, and "progress" to ancient provincial cities like Marseille, they also carried with them certain attitudes toward the environmental settings in which the cities were located. Believing that the air was literally soaking up the filth from urban streets and circulating it directly into citizens' porous bodies through their breathing, proactive doctors and public health officials created an ambitious plan for urban reconstruction. Designed to combat the dangers of stagnant, choleric air, the new urban plan for Marseille—with broad, ventilated streets—welcomed the northwesterly mistral's purifying gusts into the city's physical structure.

Urban Planning, the Mistral Windscape, and the Making of Healthy Air in Marseille

Because of the widely held belief that cholera miasma emanated from unclean places, the struggle against the disease grew to involve not only medical experts but also an emerging public health infrastructure that relied on a range of French agencies and actors, including municipal governments, sanitation services, and urban planners. In the case of Marseille, the recent cholera epidemics shone a light on problems with city infrastructure, overcrowding, and waste that had already spiraled out of control. By the mid-nineteenth century, the medieval city of the past, with its tightly bounded walls, windmills that turned on its hills, and lone major port in the old city center, was bursting at the seams. Booming maritime commerce, coupled with new soap and textile factories, attracted thousands of new workers from rural areas and immigrants from abroad. Between 1800 and 1866, the population of the city had ballooned from 111,000 to nearly 300,000.[53] With no plumbing or organized sanitation system, urban residents dumped human and animal waste into streets, waste that gravity carried down into the port below. From the perspective of Hippocrates's disease ecology, Marseille was "a nightmare of urban sanitation" whose foul air was ripe for the next cholera outbreak.[54]

In France, the model for clean-air-based urban renewal originated with French Emperor Napoleon III's chief urban planner, the Baron Haussmann, who undertook a massive street-clearing project in Paris during the 1850s and 1860s. Known for his brutal approach to dealing with "unhealthy" urban neighborhoods—regardless of their role in history or their sentimental meaning to poor communities—Haussmann mercilessly sliced through most of Paris's old medieval neighborhoods, with their zigzagging streets and their pools of refuse thought to cause tuberculosis and cholera, replacing them with broad and airy "hygienic avenues." Looking to Paris as their model, Marseille's progress-minded urban planners argued that the key to creating a safe, healthy, breathable urban atmosphere was to open up their city's structure so that air could freely move.

With the approval of Napoleon III, municipal authorities in Marseille adopted Haussmann's methods and began clearing away many of the city's overcrowded and dingy older neighborhoods during the late 1850s and early 1860s. "Except in rare spots where air circulates and renews itself, [in places with] large concentrations of men, crammed into narrow streets, where one dumps waste, where water stagnates . . . the air [in Marseille] does not refresh itself," explained one city health official, Dr. Charles Guès.[55] Replacing Marseille's tangle of medieval neighborhoods with "large, ventilated streets," like the rue Impériale, would not only open up a vital new commercial artery from the upper city down to the main port; it would also prevent stagnant water from accumulating.[56] Together with broader city streets, new green spaces like squares and parks offered refreshing air while the Canal du Midi and a new sewage system ensured the cleanliness of city water. All these efforts at urban renewal were designed to halt the accumulation of stagnant water that could harbor the cholera miasma in its vapors.

But how well could an urban plan for "ventilated streets" originating in Paris fit into the particular environment and climate of the South of France? Even as demolition teams were tearing down parts of Marseille's old city, some commentators noted that Haussmann's model of broad, open avenues was not a good fit for a windy city like Marseille, where the mistral blew for an average of 176 days per year.[57] Whereas the old medieval city consisted mostly of dilapidated buildings oriented toward the northwest, where they were angled to block and divert the mistral, the new city streets were not designed with an awareness of their vulnerability to dangerous gusts of wind. The problems associated with Haussmann-style avenues are now a recognized result of the so-called Venturi effect, meaning that wind accelerates naturally when it is canalized.[58] "In the winter, when the mistral blows, the houses on the rue Impériale are almost completely uninhabitable," noted one article on the wind's impact.[59] Even the best-heated modern apartments were freezing, and it was hard to keep the modern lampposts in the streets illuminated when the mistral was blowing.

One contemporary scientist noted that Marseille's new urban plan—which had been imported from Paris rather than designed to fit the distinctive climate and topography of the southern city—had caused *more* disease rather than made the city healthier. "Ever since the flattening of the Lazaret Hill that protected part of Marseille from the mistral and the opening of the rue Impériale, we have observed an increase in colds, and catarrhs and chest ailments have become commonplace . . . Progress does not consist in killing people," he lamented.[60] But others viewed such minor ailments as a small price to pay compared to the overall benefits that the city's increased exposure to the mistral had brought. "The mistral," wrote Bruno Martin in his 1866 encyclopedia of Marseille, "which is sometimes decried as an irritating, cold, and tiring wind, far from being considered unclean and unhealthy, is, to the contrary, a true benefit for all the regions that it visits." Marseille, he claimed, "owes [to the mistral] the cleansing of its ports and its streets, the purity of its atmosphere, and a healthy and advantageous influence on the constitution of its inhabitants."[61]

Now more than ever exposed to the mistral following Marseille's urban facelift, city residents sometimes enjoyed poking fun at their windswept lives. One humorous postcard, picturing one of the new city neighborhoods, showed the towers of the Church of St. Vincent de Paul, located at the intersection of recently completed ventilated streets, wobbling in the mistral (see figure 4.4). Below, city residents looked like tiny ants trying to make their way through the broad open avenues with their fashionable four-story apartment buildings. Ironically, by the time that

Figure 4.4. During its urban revitalization in the second half of the nineteenth century, Marseille's streets expanded to let the wind come through and cleanse the city of dirty air. As a result, Marseille became even harder to walk through when the mistral blew. "Le Mistral à Marseille. Église St-Vincent-de-Paul." Archives de Marseille.

most of Marseille's new urban plan was complete, the environmentally driven explanation for cholera was largely discredited by germ theory. Thanks to scientific discoveries by John Snow, Louis Pasteur, and Robert Koch, it became clear that cholera was indeed carried by *bacillus vibrio cholerae*, which thrives in humid places, particularly water, and enters the human digestive tract by mouth, infected hands, or tainted foods.[62] But even as the perceived threat of airborne cholera receded from Europe in the late nineteenth century, the problem of anthropogenic air pollution was growing increasingly serious.

Manmade Miasma: Combating Industrial Pollution with the Mistral

Whereas the deadly particles that French doctors had visualized in Marseille's hazy skies during a cholera outbreak turned out to be imaginary, the human-made pollution emanating from the city's steamships and factories created a sickly cloud of smoke and vapor that was all too real.[63] Late nineteenth-century Marseille was home to a number of industries that all affected air quality: factories and warehouses for fertilizer and manure, pigsties, boilers and steam engines, factories for the production of iron and leather, sugar refining, lime kilns, iron smelting, distilleries, paper production, chemical mixing, warehouses for charcoal, as well as warehouses for carcasses, dies, ink, soap, and pottery.[64] Prominent local businesses included soap and sugar refineries, tanneries, leather producers, distilleries, oil mills, cotton and wool manufacturers, and foundries.[65] One city guide remarked that many people dreamed of taking holidays from Marseille to the countryside in order to "breathe something other than the nauseating emanations of the swatches of colonial merchandise or the soap, oil, or refinery products that often form the atmosphere of the entire lower city."[66]

For some, the threat of air pollution, like the threat of cholera before it, was the natural result of social ills. One of the most famous commentaries on European air pollution came from John Ruskin, a well-to-do writer and social critic who moved from urban Britain to the rural Lake District, eager to flee the degraded environment and society of industrial modernity. In his famous public lectures from 1884, which he called *The Storm-Cloud of the Nineteenth Century*, Ruskin painted a foreboding picture of a new climatological phenomenon, which he labeled variously as a "storm-cloud," "plague cloud," or "plague wind."[67] The amorphous cloud, observable across all of Europe, was "malignant" and "degrading." Eschewing measurement-obsessed modern meteorology, Ruskin favored qualitative descriptions of the cloud. For him, the cloud and its "calamitous wind" was "a physical and spiritual manifestation of the corruption of human desire" and the sins of capitalism and materialism that urbanization encouraged.[68]

In France, as in Britain, government officials tackled the "calamitous wind" of air pollution in a piecemeal and uneven fashion.[69] On paper at least, Marseille was supposed to be solving its air quality issues by requiring industries to monitor pollution at the source since the early nineteenth century. As far back as 1826, the Marseille Public Health Council approved a techno-fix solution to industrial pollution in the form of a Rougier condenser which the prefect of the Bouches-du-Rhône department, Christophe de Villeneuve, inspected himself.[70] The condenser employed a piece of litmus paper, which, upon turning red, would alert an inspector that hydrochloric acid was present. Lead factories in Marseille also received condensers, first installed in 1851, which were used to "remove and condense fumes laden with soot, sulfur dioxide, metallic and metalloid particles."[71] As Xavier Daumalin has argued, "In both the soda and lead industries, condensers were praised by the prefectural administration as a miraculous answer to pollution and environmental conservation issues—a panacea that would ensure the continued operation of these industries."[72] Daumalin notes that some factory owners suggested that the mistral could also be used as a "natural cleanser" of industrial smoke. Plans for lead factories in Marseille often included vertical chimneys to direct pollution into the upper atmosphere, where the mistral could conveniently sweep any unwanted particulates out to sea.[73]

Just as Marseille's factory owners framed the mistral as an ecosystems service that naturally cleansed the air of pollution, locally based engineers promoted schemes for using the mistral's sky-sweeping force as a natural remedy for the soiled waters in the port. Because Marseille was built on conical and sloping surfaces, all city streets lead down to the coastal water. As a result, the Port of Marseille was the recipient of much of the city's waste, becoming "an infected cesspit."[74] One study on the "radical sanitation" of the Port of Marseille argued that the mistral provided a natural solution to the industrial problem. "When the sea is agitated by the mistral," argued the study's author, "the products of the drains will be naturally pushed to the coast."[75] Another study, by Rodolphe Serre, proposed a scheme for "the renewal of the water in the Port of Marseille in thirty-six hours."[76] Knowing that the natural force of the mistral was capable of creating two-meter-high waves, he proposed to build a canal the same length as one of Marseille's coastal jetties that could capture water. He calculated that thanks to his engineering work, the Old Port of Marseille would be entirely filled with clean water at almost no cost, since "nature happily undertakes the execution of all the work."[77] If its natural setting were used properly, he argued, Marseille could "become the healthiest city in the world."[78]

Conclusion

Though the miasma theory of disease transmission was proven wrong by the end of the nineteenth century, the powerful ideas about "good" and "bad" air that physicians had instilled among the public in Provence had lasting effects. In seeking out the causes of plagues everywhere from the remote coastal wetlands to polluted urban ports, medical experts in Provence framed local and regional-scale environments as critical spaces for securing the physical well-being of French citizens. From the perspective of nineteenth-century public health officials, the mistral's presence in the South of France was a blessing rather than a burden: the wind's sky-sweeping gusts were thought to protect the local population from the worst ravages of airborne epidemic diseases. As misguided as the miasma theory was, it gave rise to a laudatory discourse about the mistral that strengthened the Provençal regionalists' mission of carving out a distinctive subnational identity for their people.

The relationship between human bodies and provincial environments not only fascinated physicians in nineteenth-century Provence but also attracted great interest among visual artists. For a new generation of landscape painters flocking to the South of France, the region's windswept landscapes and seascapes became outdoor laboratories for testing out innovative painting materials and techniques. Through the practice of *plein air*, or "open air," painting, European landscapists situated their bodies directly in the path of the mistral. They became riveted by the northwesterly wind's momentary visual effects—cloudless blue skies, swaying cypresses, and churning sea waves—as well as its lasting climate impact in the forms of aridity and erosion. For modern painters, the mistral's role as southern France's "sky sweeper" mattered not because it cleared miasma particles away, but because it shaped the authentic look and feel of a distinctive European regional landscape. It is to these landscape painters, and the physical and emotional connections that they found with the mistral, that we now turn.

CHAPTER 5

A Sense of Place

PAINTING THE MISTRAL IN THE OPEN AIR

> What a funny thing the *touch* is, the brushstroke. Out of doors, exposed to the wind, the sun, people's curiosity, one works as one can, one fills one's canvas regardless. Yet then one catches the true and the essential—that's the most difficult thing.
>
> VINCENT VAN GOGH

Mixed in with the ocher, cobalt, cadmium, and ultramarine-blue swaths of color that grace the surfaces of nineteenth-century Provençal landscape paintings are tiny specimens of nature. The organic mementos of the southern French environment—leaves, seeds, sand, dirt, pebbles, and even insects—were carried in the wind and landed onto artists' canvases, where they were infused with brightly hued chemical paints, becoming part of the final works of art.[1] The presence of these windblown natural artifacts in the world-famous oeuvres of Claude Monet, Vincent van Gogh, and Paul Cézanne, among others, can be attributed to a new method of landscape painting that rose to prominence in nineteenth-century Europe. Rather than confine themselves to the stuffy atmospheres of traditional indoor ateliers, plein air landscape painters ventured outside, where they immersed themselves directly in real-world environments. Their insistence on painting outdoors, exposed to the elements, not only fundamentally transformed the history of European art, but it also revolutionized popular attitudes toward nature in the industrial age.

While the French administration in Paris was hard at work taming and "civilizing" its varied provincial environments in the name of modern progress, plein air painters found inspiration in the natural diversity of climates and ecosystems that existed across the country. Their close-up, intimate observations of outdoor settings echoed the work of nineteenth-century French physicians who studied place-based and seasonal influences on the human body. To create their art in locations ranging from gnarled olive groves to chalky seaside cliffs, landscape painters equipped themselves with tools specifically designed for the open air, including lightweight foldable easels, portable tubes of chemical paints, and painting umbrellas.[2] By immersing their bodies in distinctive regional weather systems, terrains, and ecosystems, plein air painters gave the French public a fresh, modern way of relating to their landscape. In contrast to the uniform and tightly controlled images of territory found on state-produced maps from the nineteenth century, modern artists left behind a civilian-produced archive of territorial images that captured the living three-dimensional reality of France as a palpable, motion-filled, and heterogeneous collection of places.

For plein air landscape painters seeking novel and stimulating sites for their outdoor studio work, Provence quickly emerged as the environmental laboratory of choice. Perceived as one of the wildest and most undeveloped regions of France, Provence, like its sister province of Brittany in the northwest, attracted artists eager to experience the authentic look and feel of rural landscapes unmarred by large-scale industry. Thanks to recently built national and international railway networks, avant-garde artists from cloud-covered northern climates flocked to the South of France, searching for immersive experiences in an exotic and sun-drenched Mediterranean environment. For many of these newcomers expecting to find a tranquil, Eden-like natural setting, the mistral came as a rude awakening. The awesome natural power of its unrelenting and violent gusts presented plein air painters with both practical and conceptual challenges that they responded to with remarkable resilience and creativity.

This chapter examines the work of painters from three major nineteenth-century artistic movements—the École de Marseille, the Impressionists, and the Postimpressionists—who painted Provence in the open air, creating innovative narratives about the human-wind relationship in the process. For homegrown painters from the École de Marseille, a regional school of Naturalist art aligned with the Félibrige group of writers, the mistral's material power was primarily visible in its longue durée climatic effects. Their ocher-hued canvases presented their "little homeland" (*petite patrie*) as an arid working landscape, where hardscrabble shepherds, cattlemen, and washerwomen labored valiantly in a dusty and timeless environment.[3] On the other hand, for Impressionists and Postimpressionists working in Provence, many of whom were only visiting the region for short stints, the mistral held a different sort of region-making power than it did for locals. In paintings by Monet, originally from Normandy, and van Gogh, a Dutchman, it was the mistral's immediate sensory effects—its dramatic impacts on the sky, the sea, cypress trees, and human bodies—that gave their viewers a visceral feeling of place. Balancing the perspectives of regionalist and cosmopolitan groups of painters in Provence was Cézanne, who transported ideas from the avant-garde Parisian art world back to his native Aix-en-Provence, where he captured the solidity of wind-hewed landscapes and the enduring presence of wind-adapted rural buildings surrounding his studio from a groundbreaking visual perspective.

By focusing on small-scale outdoor settings where gusts of wind were free to move about in an uncontrolled and spontaneous manner, plein air artists challenged the idea of human mastery over nature that had guided French state policy toward regional environments for much of the nineteenth century. The growing popularity of their affective, site-responsive works of art signaled the rise of an unofficial regional consciousness in modern France that was entwined with a bottom-up civilian movement to appreciate, tour, and preserve nature in the industrial age. In framing Provence as an environmentally distinctive part of France, plein air painters expanded the very concept of regional identity beyond its traditional linguistic, culinary, and ethnographic definitions. Fanning out across the region's beautifully rugged physical landscapes, plein air painters discovered a sense of Provençal-ness in the nature all around them, including in the air itself.

Nativist Nature: The École de Marseille and the Origins of Plein Air Painting in Provence

It was during the first several decades of the nineteenth century that European painters, rejecting the confines of Enlightenment rationality, turned in large numbers to nature's wild and unbridled power as their muse. These Romantic-era artists, such as Théodore Géricault and Eugène Délacroix in France, Caspar David Friedrich in Germany, and J. M. W. Turner in England, conceived of nature as a powerful and autonomous force that dominated human life. Disavowing the Enlightenment philosophers' confidence in humanity's power to know and ultimately control nature through scientific study, the Romantics focused instead on the mysterious, the sublime, and the semisacred human experience of nature's might. Their canvases pictured landscapes and seascapes stirred by violent weather, often to the peril of a traveler or a sailor facing immediate danger or death. The visual trope of a gust of wind, from the perspective of Romantic painters, was simultaneously an expression of nature's power and a signal that something dramatic was about to happen.[4] Featuring rustling leaves, moving garments, and tumultuous skies, their windswept scenes were intended to serve as "a reflection of the torments of the soul."[5]

Despite Romantic painters' interest in the materiality and the emotional symbolism of wind, the reality was that their oeuvres were born from the safe confines of indoor studios, or ateliers. A completed canvas often reflected the work of an entire team of apprentice painters rather than a firsthand impression of windblown nature from a single artist. It was only during the 1820s and 1830s that European landscape art shifted its emphasis away from the artifice of academic painting and toward a more honest depiction of nature based on direct observations. Naturalism, as this more "authentic" approach to landscape painting was called, aimed to rid the art world of its imaginary representations of nature and instead bring artists into actual physical contact with it. In France, Naturalism got its start through the Barbizon School, a group of artists who painted outdoors in the forest of Fontainebleau, outside Paris. This new art school, which got its name from the small village in the forest where the artists lived, was part of a growing bottom-up movement of French citizens that wanted to protect idyllic rural areas from the claws of an expanding industrial economy. For Barbizon artists, open-air painting in their tranquil forest was a means of elevating nature as a public good and counteracting the juggernaut of modernization that was threatening to overwhelm one of Paris's last remaining green spaces.[6]

Naturalist painting found its way to Provence through a young artist named Émile Loubon, a native of Aix-en-Provence with an interest in landscape art. In 1832, at age twenty-three, Loubon made the same trek that provincial artists in France had made for generations, traveling north to Paris for training.[7] During his time in Paris, Loubon grew tired of the classical ateliers of the city's formal art schools, deciding instead to join the young group of painters at the Barbizon School. Loubon spent years as an artist-in-training under the tutelage of the Barbizon circle's founding members, including Théodore Rousseau and Narcisse Diaz de la Peña. Together, the artists spent days outdoors in the forest of Fontainebleau, painting in the wind and the rain, and observing oak trees, cloud-heavy skies, gullies, streams, and rural laborers cultivating and harvesting the land.[8]

Loubon's serendipitous encounter with the Barbizon School convinced him that the physical environment itself offered a powerful means of anchoring one's sense of self and one's sense of local community at a time of national-scale territorial consolidation and industrial change. Yet even as he supported the Barbizon School's goals of preserving the forest of Fontainebleau from modern development, Loubon felt out of place in northern France. The artist found himself thinking back to the starkly different French territory of his childhood: the dry earth, the swaying cypresses, and the bright light of inland Provence. Like the inhabitants of Fontainebleau, people living in Provence's rural towns and villages were facing rapid economic and cultural changes that threatened to upend their traditional livelihoods. Could the Naturalist philosophy and techniques that he learned from his mentors at the Barbizon School help to valorize and protect the rural landscapes of his native region?

In 1845, at the age of thirty-six, Loubon would get his chance to explore the possibilities of Naturalist painting in Provence when he was invited to serve as the new director of Marseille's leading art school. The École des Beaux-Arts de Marseille was one of several regional art schools established in France during the mid-nineteenth century. Such schools aimed to provide quality training to provincial French painters without requiring them to move to Paris. Because they opened their doors at precisely the same moment that plein air landscape painting was becoming fashionable, these schools became important centers for the observation, recording, and dissemination of site-specific provincial nature imagery. As institutions of learning, they became incubators for a new, decentralized discourse on French nature rooted in what ecologists would now call bioregions: territories defined by their unique blend of geological, ecological, and climatic features.

Bringing a fresh style of rustic Naturalism to the South of France from his training in the forests outside Paris, Loubon fostered a new artistic movement in the Midi, a sort of "Barbizon Provençale,"[9] that encouraged young students to paint nature in a radically new way, based on their direct observation of their provincial surroundings. Loubon required his pupils at the École des Beaux-Arts de Marseille to leave the atelier and to explore the world outside, bringing their canvases, easels, and palettes with them. It was through their physical encounters with nature, he hoped, that his students would capture the distinctive ecological character of Provence, "the special alchemy of sunlight, wind, and humidity that produced a sense of place."[10]

A core element of the Provençal ecosystem, the mistral was a frequent presence in the works of Loubon and his circle of painters. Keeping in line with the Barbizon School's view of the natural world as a humanized place where people were almost always present, Loubon's paintings often featured groups of rural laborers going about their daily work in a countryside alive with movement. His work conveyed a feeling of Provence as a land that was "hard, severe, buffeted by the wind," and he preferred to paint figures in a way that "gave the impression of a struggle against animals and the elements."[11] In his paintings, Loubon demonstrated a keen awareness of the mistral's transformative power over the southern atmosphere and southern light. As Provence's *balayeur du ciel*, the mistral rid the sky of clouds, making it possible to create an exceptionally clear canvas of bright, sunlit colors with each rock and cliff face outlined in bold starkness. Scientific research has since proven the wind-driven effects on atmospheric clarity that Loubon observed, establishing a correlation between the blueness of the sky and the presence of wind.[12]

Both the short-term weather effects and long-term climate impacts of the mistral are visible in one of Loubon's most famous canvases, *View of Marseille from the Aygalades on a Market Day*, completed in 1853 (see plate 4). The clear sky, the smoke blowing from Marseille in the distance, and the crisp outlines of the gray limestone cliffs of the Calanques to the east of the city all demonstrate the short-term atmospheric effects of the wind. But besides its immediate impact on the sky's color and clarity, Loubon also takes note of the mistral's long-term climate impacts on the physical landscape: the arid ground that lacks vegetation, the dust visible around the cattle, and the wind-bent tree in the middle-right of the canvas. As art historian Greg M. Thomas has argued, Naturalist painters were among the first truly ecological thinkers in nineteenth-century Europe, even predating the work of scientists such as Darwin and Haeckel.[13] Their works of art expressed what one might call "earth narratives," which Thomas defines as "the many interconnected organic processes of natural growth and movement."[14] The mistral, interpreted through this ecological lens, was not simply an ephemeral or momentary phenomenon; it operated, rather, on an expansive chronological timescale that gave it the power to slowly erode, desiccate, and sculpt the land. In highlighting the slow climatic work of the mistral through his painting's parched, earth-toned foreground, Loubon established a narrative of Provençal nature as a well-worn, ancient landscape that he revered as a thing of beauty.

But it was not just the organic ecological processes taking place in the scene—the invisible exchanges among earth, plants, and wind—that drew Loubon's attention. Like his mentors at the Barbizon School, Loubon wanted to highlight the close bonds that existed between human beings and their natural environment. In focusing on the physical work of a group of Provençal cattle drivers struggling to move their animals through an arid, windblown landscape, Loubon produced an admiring portrait of pastoral labor in Provence. As Caroline Ford has argued, for nineteenth-century bourgeois painters who were tired of revolutionary disorder, the peasantry "became nostalgically romanticized as the embodiment of order and social stability."[15] In *View of Marseille*, Loubon created an iconic rendering of rural France that elevated the common agricultural laborer to heroic status. Meanwhile, the natural environment itself took on a mythic and ancient quality, defying the creeping realities and uncertainties of industrial modernity visible in the distance.

While many of Loubon's paintings referenced the physical challenges faced by ordinary rural laborers in mid-century Provence, they also reflected his privileged place in French society. Not only did his canvases capture the exhausting work of animal husbandry from the vantage point of a comfortable bourgeois artist, but his practice of outdoor painting was only possible because of his fortunate position as a male landscapist. Because plein air artists needed to carry their easels, canvases, palettes, and tubes of paint through rugged and sometimes inhospitable landscapes, nineteenth-century open-air painting was de facto a "woman-free zone."[16] Indeed, there were no women among the 378 artists enrolled as students at the École des Beaux-Arts de Marseille.[17] Yet even as they were excluded from work as landscape artists, Provençal women, particularly peasant women, became a popular subject matter for male Naturalists.

In his 1857 painting *The Washerwomen*, for example, Loubon pictures a group of women struggling to hang up and fold sheets as the mistral is blowing (see figure 5.1). Two of the women are standing and hanging the sheets, their anonymous faces hidden from view entirely. Only the face of the woman seated in the foreground is shown. In her attempt to fold the laundry, she has placed several sheets onto the ground and weighed them down with rocks. Not unlike Loubon's painting of the male cattle drivers, this image highlights simultaneously the mistral's short-term weather effects and its long-term climatic impacts on the landscape. The foreground features a stony hilltop terrain, hewed over many seasons by the wind, while the cloudless sky is colored deep blue. Meanwhile, in this harsh yet beautiful Provençal environment, hardworking local washerwomen calmly persevere in their daily work as their clean white sheets are nearly torn from the line.

In encouraging his students to observe and record the special relationship between Provençal people and their arid working landscape, Loubon helped to popularize the regionalist idea of Provence as a distinct homeland (*petite patrie*) within the larger French nation (*grande patrie*). Like the Félibrige group of poets and writers, founded by Frédéric Mistral in Arles during the same mid-century period, many of the painters affiliated with the École de Marseille wanted to preserve the Provence of yesteryear from economic transformation and Parisian-led cultural homogenization. While committed to painting from first-hand observations, their canvases nonetheless privileged seeing certain things—traditional pastoral laborers immersed in rural landscapes—

Figure 5.1. Émile Loubon, *The Washerwomen*, 1857. Oil on wood. Fonds d'oeuvres de la Région Provence-Alpes-Côte d'Azur, dépôt d'Arsud l'outil des Arts et du Spectacle, au Musée ZIEM. Photo by Robert Terzian.

while omitting other realities, such as industrial workers in growing urban centers like Marseille. Theirs was an idealized vision of a timeless Provençal society that existed organically in tandem with its arid, windswept climate.

Nature in Motion: Claude Monet and the Atmosphere of Antibes

While local painters affiliated with the École de Marseille focused on their region's arid working landscapes, artists who came to Provence from northern France and other European countries produced different kinds of influential regional images that reached national and international audiences. Many of these newcomers to Provence were city dwellers, tired of the trappings of modern life, who looked to the southern Mediterranean environment as an attractive reprieve from industrial civilization. Fashionable tourist guidebooks even likened Provence's colorful scenery and abundant sun to the pristine and exotic landscapes of colonial Africa.[18] These promotional tourist materials, however, did little to prepare visitors for the violent force of the mistral. In contrast to homegrown painters, who were intimately familiar with their regional wind, the Impressionist painters who flocked to Provence during the 1870s and 1880s were often shocked when the mistral arrived uninvited to their plein air painting sessions. Despite the frustrations it caused, however, the mistral's fugitive and fast-moving presence offered visiting painter-tourists a unique opportunity to study and capture the ephemeral, shifting, and contingent aspects of life in the South of France.

Impressionism, a loosely defined art movement whose name was first coined in 1874, was deeply concerned with the ever-present movement of the natural world. "Where an academic painter sees white light, the Impressionist sees a thousand colors," explained one prominent art critic. "Irregular touches animate the landscape, giving it life. Tones and touches give a variety of atmospheric states, with each plane not immobile, but shifting."[19] Notably, the Impressionists' interest in the weather grew in tandem with the emergence of meteorology, which, as I discussed in chapter 3, came into its own as a discipline and as a centralized government bureau in France during the 1870s and 1880s. "The artists' fascination with atmospheric effects and weather," art historian Richard Brettell has argued, "was so important to their aesthetic that one would not be criticized for linking them to contemporary meteorologists who were then actively studying the movement of clouds, winds, water, etc., across the surface of the earth throughout the year and the day."[20]

Among the most famous Impressionists to visit the South of France in the late nineteenth century was Claude Monet, the native of cloud-covered Normandy who, like many of his contemporaries, was lured by popular images of Provence as a sunny paradise. In January 1888, during the depths of winter in the North of France, Monet took a luxury train from Paris to Marseille and then traveled eastward along the Mediterranean coast to the fortified seaside town of Antibes.[21] After settling into his hotel room at the Château de Pinède, Monet explored his coastal surroundings on foot, often carrying his painting supplies with him. It was through his direct bodily and sensory contact with the Provençal shoreline that Monet experienced the legendary mistral firsthand. "Curses and despair!" he declared in a letter to his wife, Anne. "I returned chased by the wind; I wanted to work all the same, securing the canvas and the easel, but my palette and my canvas were covered with sand; I had to give up."[22]

Yet even as he complained, Monet was captivated by the mistral's transformative effects on the Provençal coastline. Thanks to his quick, energetic brushstrokes in his 1888 painting *Cap d'Antibes, Mistral,* we can almost feel the wind blowing across the Mediterranean Sea, giving us a visceral, sensory impression of the mistral's relentless force (see plate 5).[23] What comes across most prominently in Monet's canvas is the enormous sky that dominates much of the painting's composition and is, to a significant extent, the driving medium of the entire landscape.[24] In allowing the sky to take up such a generous portion of the image, Monet was able to give the viewer a detailed understanding of the dynamic windblown atmosphere that obscures the southern Alps that appear faintly on the other side of the cape. Besides the disturbances that the mistral brought to the sky, Monet observed the wind's transformative effects on the other natural elements in the scene: the trees that swayed in the foreground and the sea that became choppy and dark. In one of his letters, he noted how the wind changed the colors of the world around it, producing "a sea animated by a multitude of accents of periwinkle, indigo, coral, greens, yellows."[25]

Emphasizing the mistral's energic movement across the sky, the Mediterranean Sea, and the coastal vegetation, Monet's painting reflected the viewpoint of a sensitive modern observer recording his novel experience of Provençal weather at a particular moment in time on a particular day of the year. It expressed what art historian Anthea Callen calls the Impressionists' "visual culture of immediacy": their desire to highlight the very feelings of nonstop motion and contingency that characterize modern life.[26] Through his quick brushstrokes and dashes of color, the painter himself collapsed into the scene, his body completely immersed in the power of the organic processes going on around him.[27] In contrast to regional painters like Loubon, Monet was not interested in depicting Provence as a traditional working landscape that supported the livelihoods of its people. Rather, the landscape that appears in *Cap d'Antibes, Mistral* is the creation of an outsider artist with no interest in the longue durée climate or historicity of the landscape.

Summing up their aesthetic goals, art critic Jules-Antoine Castagnary once argued that Impressionists "render not the landscape but the sensation produced by the landscape."[28] Favorite Impressionist terms like *effet, sensation,* and *impression* underscored the significance of the painter's subjective bodily experience to the visual image that took shape on their canvas.[29] Yet it was precisely the Impressionists' focus on recording nature's fleeting sensory stimuli that left them vulnerable to criticism from other modern artists of their time. During the 1880s, an emerging group of European painters cultivated a new method of landscape representation that still relied on plein air techniques but revealed more than the artist's momentary, surface-level impressions of their environment. Known as the Postimpressionists, these avant-garde painters exposed the deeper—and often unseen—symbolic, cultural, historical, and emotional meanings carried in physical landscapes.

Postimpressionist Travelers to Provence: Paul Gauguin and Vincent van Gogh

In 1888, during the same year that Claude Monet painted the windswept Mediterranean coastline in Antibes, French artist Paul Gauguin and his Dutch friend Vincent van Gogh moved to the ancient Provençal city of Arles, where they hoped to found a new artistic community called the "studio of the South."[30] Like many of the northern European artists who traveled to Provence before them, Gauguin and van Gogh were prepared to encounter a warm and gentle climate, only to find themselves confronted, for large stretches of time, by the chilly gusts of the mistral. As they reluctantly adapted to the mistral's formidable presence, the two visiting artists became intrigued by the wind's invisible entanglements with local landscapes and communities. During their brief but productive stays in Provence—which included van Gogh's relocation from Arles to a mental institution in nearby Saint-Rémy—the artists applied their novel Postimpressionist ideas to the mistral's material and symbolic power.

For a global traveler like Gauguin, Provence offered an enticing opportunity to paint people who were citizens of modern France but who spoke, dressed, ate, and lived according to a set of regional traditions that he considered exotic. During the late nineteenth century, it was not uncommon for men like Gauguin to flee their metropolitan homes for what they saw as a simpler life in far-flung French provinces or in overseas European colonies. Once outside their cities, these men became ethnographers: modern European observers who undertook the systematic study of "primitive" or "traditional" cultures that were supposedly unblemished by the corruption of modernity.[31] One of the chief ways in which Gauguin developed his ideas about primitivism was through landscape paintings featuring women that he encountered in different parts of the world.

Not unlike his earlier tableaux of women in white bonnets tending the fields in rural Brittany, or his later paintings of Polynesian women on the beach in the South Pacific, Gauguin's *Arlésiennes (Mistral)* (1888) frames a group of women from Arles as living embodiments of Provence's traditional premodern culture (see plate 6).[32] Inspired by an actual group of elderly women that he spotted outside the window of the little yellow house that he shared with van Gogh, the painting depicts four figures in traditional Provençal coiffures and thick cloaks making their way through the city's Place Lamartine as they struggle against the force of the mistral. Cone-shaped yellow coverings over the square's trees suggest that it is wintertime, which would have deepened the mistral's chilling effects on their bodies. In spite of their physical discomfort, the huddled forms, emotionless eyes, and covered mouths of the two women in the painting's foreground revealed their calm and obstinate resignation to their local wind as they process forward.

While Gauguin observed the mistral's role in traditional Provençal culture from the perspective of a nineteenth-century ethnographer, his housemate van Gogh developed his own tactile approach to painting the mistral. "I've already had an opportunity to find out what this mistral's like," he wrote to his brother Theo shortly after arriving in Arles during March of 1888. "I've been out on several hikes round about here, but that wind always made it impossible to do anything. The sky was a hard blue with a great bright sun that melted just about all the snow—but the wind was so cold and dry that it gave you goosebumps."[33] Despite its chilling effects, the mistral's ability to generate a magnificent blue sky made it possible for van Gogh to capture color contrasts in the natural environment more effectively than he could in the dreary North. It was now feasible for him to use an intense orange and an intense blue together on the same canvas.[34]

In order to paint outdoors amid the swirling gusts of the mistral, van Gogh devised his own system for hammering his easel into the ground to keep it from blowing away during his plein air painting sessions (see figure 5.2). "You shove the feet of the easel in and then you push a 50-centimetre-long iron peg in beside them," van Gogh explained in a letter to his friend Émile Bernard. "You tie everything together with ropes; that way you can work in the wind."[35] And yet, even with his easel fixed securely to the ground, van Gogh was often unable to control the movement of his hands when the wind blew. "I always have to struggle against the mistral, which absolutely prevents one being in control of one's brushstroke," he explained in a letter to his brother Theo, "Hence the 'wild' look of the studies."[36]

One of the paintings that van Gogh composed with the help of his special easel was *Cypresses with Two Figures*, an image that explored the transformative effects of the mistral on Provence's rural farming land-

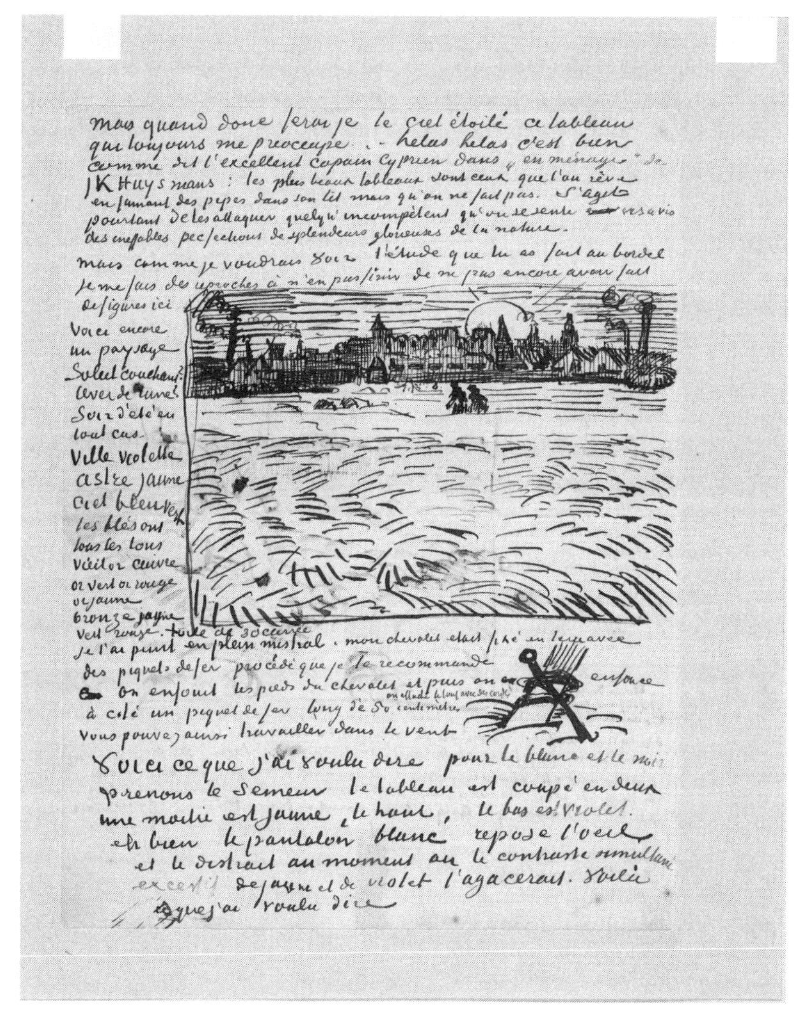

mais quand donc ferai je le ciel étoilé ce tableau
qui toujours me preoccupe . - helas helas c'est bien
comme dit l'excellent copain Cyprien dans "en ménage" de
JK Huysmans : les plus beaux tableaux sont ceux que l'on rêve
en fumant des pipes dans son lit mais qu'on ne fait pas. S'agit
pourtant de les attaquer quelqu'incompétent qu'on se sente vis a vis
des ineffables perfections des splendeurs glorieuses de la nature.
mais comme je voudrais voir l'étude que tu as fait au bordel
je me fais des reproches à n'en pas finir de ne pas encore avoir fait
de figures ici
Voici encore
un paysage
Soleil couchant?
Lever de lune?
Soir d'été en
tout cas.
Ville violette
astre jaune
ciel bleu vert
les blés ont
tous les tons
vieil or cuivre
or vert or rouge
or jaune
bronze jaune
vert rouge. toile de 30 carrée
Je l'ai peint en plein mistral. mon chevalet était fixé en terre avec
des piquets de fer procédé que je te recommande
on enfoui les pieds du chevalet et puis on enfonce
à côté un piquet de fer long de 50 centimètres. on attache le tout avec des cordes
Vous pouvez ainsi travailler dans le vent.
Voici ce que j'ai voulu dire pour le blanc et le noir
prenons le Semeur le tableau est coupé en deux
une moitié est jaune, le haut, le bas est violet.
eh bien le pantalon blanc repose l'oeil
et le distrait au moment ou le contraste simultané
excessif de jaune et de violet l'agacerait. Voilà
ce que j'ai voulu dire

Figure 5.2. Vincent van Gogh, "Wheatfield with setting sun and leg of an easel with ground spike," autograph letter signed in Arles to Émile Bernard, ca. June 19, 1888. Pen and black ink on two sheets of cream, machine-made laid paper. Thaw Collection. Pierpont Morgan Library, New York. MA 6441.7.

scape (see plate 7). The subject of the painting was a cluster of tall cypress trees blowing upward toward the sky in what van Gogh referred to in a letter as "great circulating currents of air."[37] Like Monet, van Gogh allowed the strong bodily sensations of the mistral to shape how he depicted the wind on his canvas. His swirling brushstrokes carrying thick gobs of paint gave the impression of a three-dimensional landscape that leaped out from the canvas with blustery movement and energy.[38] But besides the immediate sensory effects of the wind that it captures, *Cypresses with Two Figures* is a painting rich in cultural symbolism.

Cypress trees were one of Provence's most recognizable geographic signifiers.[39] In addition to their utility as windbreaks to protect crops, they were used by the Romans who once ruled Provence as funerary monuments, a tradition that Provençaux retained long after the powerful empire had collapsed. As Susan Alyson Stein argues in her study on van Gogh's depictions of cypresses, the Dutch artist conceived of these towering evergreens as living beings with an alluring spiritual and historical authority. Few motifs were more emblematic of Provence or resonated more strongly with "his belief in the consoling and uplifting power of art and nature."[40] *Cypresses with Two Figures* not only accounted for the fleeting sensory effects of the undulating movement caused by the mistral, but the painting also valorized the intangible and eternal power of cypresses as ancient human adaptations to a windswept land. When activated by the mistral's energy, these living geographic signposts of Provence commanded great power, taking on a "flamelike" vitality.[41]

Van Gogh began work on *Cypresses with Two Figures* in 1889, shortly after he moved from the little yellow house that he shared with Gauguin, in Arles, to a psychiatric asylum for the well-to-do in the nearby Provençal town of Saint-Rémy. The steady mental decline that he experienced during his time in the South of France raises an important question. Were van Gogh's landscape paintings rooted in his observations of actual windswept environments, or were they more reflective of his state of internal turmoil? It is difficult to answer this question with certainty. What we do know for sure is that van Gogh spent a great deal of time outdoors during his time in Saint-Rémy—he was free to walk the grounds surrounding the asylum—and his letters suggest that he took his in situ observations of nature very seriously throughout his stay. In his correspondence with friends and family in 1888–89, he referred directly to the mistral in forty-five separate letters.[42] While many of these letters reference the wind's effects on air temperature and the appearance of the landscape, others point to the mistral's more hidden and undiagnosed effect on his mental health. "When the mistral's blowing," van Gogh lamented in one letter to his brother Theo, "it's the very opposite of a *pleasant* land here, because the mistral's really aggravating."[43] Another letter referred to "the devil of the mistral" that was with him for three-quarters of the time.[44]

Notably, the type of mental anguish that the mistral produced in van Gogh was never officially documented in the medical literature of the period in the same manner as wind-focused studies on contagious diseases like cholera. Here and there, literary accounts or published travel journals alluded to the mistral's impacts on the moods of locals and visitors. Popular travel writers like Stendhal openly described the mistral as a wind that "grates on the nerves so that the most dauntless person is unwittingly upset."[45] The few scientific articles on the mistral's mental effects in the nineteenth century tended to pile onto existing cultural stereotypes of Provençaux as barbaric and uncivilized, blaming the wind for the local population's "sudden passions, their sudden fits of anger, and their sudden returns to listlessness and languor."[46] While these sorts of informal anecdotes are plentiful, comprehensive medical research on the mental health effects of the mistral only emerged in the late twentieth century.[47] Given this lacuna, the paintings, pen-and-ink drawings, and letters that van Gogh left behind offer an imperfect yet intriguing archive for exploring the intersections of weather and mental health in the nineteenth century.

While their dream of establishing a "studio of the South" fell apart in less than a year, the visiting Postimpressionist artists Gauguin and van Gogh helped to define the popular image of Provence for national and international audiences during a golden age of late nineteenth-century rail travel. In contrast to the oeuvres produced by locally rooted artists from the École de Marseille, Postimpressionist paintings did not emphasize the region's aridity or its traditional outdoor laborers. Choosing brightly colored paints that underplayed the harsh dryness of Provence, these outsider artists invented a fresh and modern regional iconography that conveyed a feeling of discovery and exotic novelty about the southern landscape. While Cézanne, one of their fellow Postimpressionists, was also intent on capturing the unique qualities of Provençal nature, his personal connections to the region—he was born and raised in Aix-en-Provence—meant that his depictions of its windswept landscapes were fundamentally different from those of his peers.

Climate, Landscape, and Regional Heritage: Paul Cézanne's Provençal Environments

In 1886, at the age of forty-seven, Paul Cézanne—a successful Postimpressionist who had made a name for himself in the avant-garde Parisian art world—decided to leave the French capital, where he had moved in his early twenties, and return permanently to the South. His decision to fully embrace his provincial identity and set up a studio near his childhood home in Aix-en-Provence would mark a new chapter in his artistic oeuvre. Immersed once again in the dry and monumental landscapes of his beloved *petit pays*, Cézanne embraced a "style of rusticity" that blended Provençal nature and culture together with a fresh modern twist.[48] Unlike many of his fellow modern artists, who traveled to Provence as tourists, Cézanne was not interested in using the motif of the *coup de vent*, or the wind gust, as a means of conveying a sense of place in southern France. Claiming "an enormous affection for the contours of my countryside," he was far more interested in the longue durée workings of Provence's Mediterranean climate on its physical geography than on the short-term variability of the atmosphere.[49] Through his painterly lens, the mistral was transfigured from an invisible and transitory element in the sky to a slow-moving sculptor of arid, well-structured landscapes and traditional climate-adapted dwellings.

The fact that the mistral's violent and destabilizing qualities were largely muted in Cézanne's paintings did not mean that the artist found it any less frustrating to paint on a windy day than newcomer artists like Monet or van Gogh. In his biography of Cézanne, Joachim Gasquet highlighted one particularly tense episode, when the mistral demolished one of the artist's promising works in progress. Gasquet and his friend Xavier de Magallon were searching for Cézanne in the countryside near his studio in Aix-en-Provence when they came across the middle-aged man stomping his feet on the rocks, fists clenched, and crying in front of a battered canvas that had flown away in a gust of the mistral. "Flattened in the bushes, [there appeared] the touches of red on the canvas, the marble reds, the pine trees, the jeweled mountain, the intense sky . . . It was a masterpiece that equaled the nature around it," Gasquet recalled. For Cézanne, the frustration of losing his plein air painting to the whims of nature was almost too much to bear. "An enormous anger, a madness," gripped the artist as he walked toward his ruined canvas, tore it up, threw it onto the rocks, and trampled on it with his shoes. "*Foutez le camp!* [Get out of here!]," he yelled at the two men as he hid himself in a cluster of pines and wept.[50]

Despite the risk of a ruined canvas and a wasted day of work, Cézanne considered painting outdoors, immersed in the natural elements of his native Provence, critical to the evolution of his oeuvre. "Throughout the days of rain, mistral, and sun, he returned to all the routes of his youth, climbing the trails," recalled Gasquet.[51] Rather than emphasizing the restless, mobile qualities of the landscapes that he encountered, Cézanne fixated on the solid, almost timeless, quality of their forms. This stability was evident to him in both the deep geological history of the region's mountains, particularly the nearby Mont Saint-Victoire, and the enduring presence of the traditional human-made dwellings that dotted the arid countryside. Built from locally available materials and specially adapted to withstand the mistral, vernacular works of architecture such as farmhouses, huts, and pigeon towers embodied the kind of deep symbolic meaning for Cézanne that cypress trees had for van Gogh.

Houses in Provence: The Riaux Valley near L'Estaque, completed sometime between 1879 and 1882, is a striking example of Cézanne's early house paintings (see plate 2). The canvas features two *cabanons*, or small cottages, set amid a rocky hillside not far from his studio in Aix-en-Provence. Once belonging to the well-known Provençal sculptor Pierre Puget, these wind-adapted, earth-toned structures suggest a synergy between the natural landscape and human lifeways. "The tiered layout of the house," notes the art historian Nina Maria Athanassoglou-Kallmyer, "echoes the rocky layers of the landscape . . . The stony setting of the land has incorporated Puget's abandoned petrified hut, symbolically implying the great artist's indestructible bond to his native land."[52] While Cézanne's tough and stony landscape echoed the rugged images of Provençal nature found in the earlier works of École de Marseille artists like Émile Loubon, his painting was infused with a provocative modernist sensibility. Its volumes were "ample, bulky but flat, their contours and surfaces oddly smooth as if polished by man's hand and by centuries of rain and wind."[53] Later emulated by cubists Georges Braque and Pablo Picasso, his landscape was at once ecological and geometrical.

The harmonious relationship between traditional rural buildings and Provence's arid environment can likewise be found in *House in Provence*, a painting that Cézanne composed several years later, in 1885 (see figure 5.3). In this image, the viewer encounters a large mas farmhouse with a windowless wall to defend against the mistral, situated between the brown earth and twisted trees in the foreground and the gray-blue mountains in the background. In contrast to Monet's painting of windswept Antibes, the sky barely figures into the image, while the solid structural features of the landscape dominate, conveying a sense of territorial solidity and permanence. No living person is visible anywhere in the scene. Like the hilltop windmill featured in Daudet's famous short story about Provence, the empty dwellings pictured on Cézanne's canvases serve as nostalgic memorials to a once-thriving rural society that has fallen victim to the pressures of modernity.

Yet within the lifeless calm of his house paintings, a fresh vision of the Provençal countryside emerges: a preservationist's landscape. While the artist never joined a formal regionalist organization like the Félibrige, Cézanne's landscapes were unmistakably infused with a sense of pride in both the natural and the cultural heritage of Provence. Rhetorically, his images of an ancient, stable, and enduring rural Provençal countryside served as a counterweight to the fast-paced consolidation of French national territory through modern railways, industry, and cosmopolitan

Figure 5.3. Paul Cézanne, *House in Provence*, ca. 1885. Oil on canvas. Courtesy of the Indianapolis Museum of Art at Newfields.

culture. Integrating Provence's geological features, arid climate, and vernacular architecture into a seamless whole, Cézanne's works of art valorized an older set of human relations with tangible places—a slowing of time to a different, preindustrial pace. His interest in the mutual influences of place, climate, cultural heritage, and regional identity in late nineteenth-century France anticipated the growing use of geological timescales to frame environmental change in the twentieth and twenty-first centuries. Through his countryside scenes, Cézanne established a powerful philosophical connection between the stability of the natural environment and the cultural survival of his native region.

Conclusion

While each artist who painted *en plein air* in Provence visualized the region in their own way, together, their collective oeuvre suggests a growing relationship between nature and regional identity in nineteenth-century France. When homegrown painters like Émile Loubon and Paul Cézanne observed their native southern landscapes, they saw places whose geological features had been shaped, over the long term, by an arid and windy Mediterranean climate. Their paintings demonstrated how this regional climate spurred people to adapt their forms of labor and their dwellings to a territorial context that was different from the rest of France. Through their earth-toned visual images, Provençal nature and Provençal culture became symbolically intertwined, providing a novel environmental basis for region-building efforts that had often relied on narrower linguistic or ethnic frameworks. While nonlocal painters were not directly invested in defending Provence from internal colonization from Paris, their creative works played a significant role in defining the region's geographic image for national and international audiences of tourists. In highlighting the wild and untamed aspects of Provençal nature with their vibrant dashes of color, artists like Claude Monet and Vincent van Gogh promoted an idea of regional authenticity that was inseparable from Provence's blustery atmospheric conditions.

The visual evidence present in plein air Provençal landscape paintings thus complicates the notion that nineteenth-century Europeans followed an ethos of domination and control over their natural environments. While the central government in Paris and its aligned economic leadership supported policies of top-down environmental management and knowledge collection, civilian landscape painters—together with their creative counterparts in literary circles—promoted an alternative form of bottom-up nature appreciation that valorized France's rich mosaic of climatic and ecological zones. In an age of fast-paced industrialization and nation-building, the act of painting environments in situ opened painters' eyes (and the rest of their bodies) to the geographic features of France that resisted human control.

Nothing symbolized the enduring material power and autonomy of French regional environments more than the violent and overpowering gusts of the mistral. Whether it was through its slow-moving power to erode landforms and bend pine trees into twisted clawlike shapes, or through its immediate power to clear the sky of clouds or generate a sea of whitecaps, the mistral was publicly recognized, through the work of plein air painters, for its place-making power. Expanding the modern understanding of Provençal identity to include natural as well as cultural types of patrimony, modern landscape painters laid the groundwork for the kinds of region-centered sustainability initiatives that French citizens are leading today.

EPILOGUE

Living with the Mistral in the Twenty-First Century

In the ancient Provençal city of Arles, a sleek new ten-story tower with a shimmering stainless-steel surface rises prominently in the sky. Designed by the world-renowned architect Frank Gehry, the tower takes its visual cues from the natural contours of the Alpilles, the small mountain range to the city's north, as well as from Vincent van Gogh's imaginative portrayals of the region's night sky. The tower, which opened in 2021, anchors LUMA Arles, a creative complex that houses a contemporary art museum as well as workshops for emerging artists from across the globe. Surrounding the tower and its adjacent buildings is a ten-acre park conceived by Bas Smets, a Belgian landscape architect known for his forward-thinking approach to designing green spaces. Before the park could take shape, Smets had to remove the remnants of a nineteenth-century railyard that used to occupy the site. After hauling away layers of gravel, steel, and concrete slab, he was left with a tabula rasa that set the clock back to preindustrial Provence. In this freshly cleared space, he set to work on constituting the park's soil, water features, and topography from scratch.

In each period of European history, the figure of the gardener has played a critical role in mediating the relationship between nature and culture. When considering the design for his new park in Arles, Smets had an abundance of French historical models to choose from. Through its ball-shaped trees and geometrically arranged hedges, the seventeenth-century Gardens of Versailles conveyed King Louis XIV's control over both the natural world and human beings. In the nineteenth century, following the demise of the Versailles-based French monarchy, large urban parks, catering to the tastes of the fresh-air-loving French bourgeoisie, featured exotic trees, immaculate flower beds, and carefully mowed lawns. But in designing LUMA Arles's Parc des Ateliers, Smets took a radically different approach from these human-controlled versions of French garden design.

Smets's goal for the Parc des Ateliers was to create a self-sustaining public park that could grow and thrive alongside the natural processes of the Mediterranean bioregion.[1] Before he planted anything, Smets commissioned a study of the park's wind patterns (see plate 8). The study revealed how the mistral's gusts, represented by northwesterly wind arrows, would inhabit the future green spaces pictured in the center and right-hand areas of the image. Adjacent buildings would be tall enough to redirect the flow of wind in certain areas. Like the rounded apse of a Camargue hut, the curving shape of Gehry's tower, represented with a circle in the upper-left-hand corner of the plan, would split the mistral's gusts to the left and to the right of the structure.

When he kindly showed me around his completed garden, Smets explained that his goal was to work with the mistral—an essential part of the site's natural character—to create a twenty-first-century green space that would evolve over time with the long-term honing and hewing power of the wind's directional movement. In an article about the project, he put it this way:

> We wondered what nature would have done over time. We modeled how the northern mistral winds would gradually blow sediments to the site, creating a new fertile topography on top of the dead rocks. The winds would accumulate a dune-like landscape that produces a new space-time relationship. The higher the elevation of the dune, the more time would have passed to create it. A walk along the dunes becomes a journey into accumulated time, or the future.[2]

Smets, in other words, had intentionally ceded control of his park to the mistral. The green space that will take shape in Arles over the next hundred years will be the co-creation of human beings and nonhuman nature. One can already see clusters of native trees and shrubs thriving on small dunes that will grow higher over time (see figure 6.1).

While the Parc des Ateliers reflects the kind of long-term thinking needed for successful regional climate adaptation, it is also a public space that showcases the spectacular real-time visual effects of the mistral's movement. In the center of the park is a large pond, represented by a curving white form in the center of Smets's wind study. Reminiscent of the nearby Camargue wetlands, the pond features beds of reeds—like the ones that van Gogh used as pens for his outdoor studies in Provence—and provides a beautiful watery surface where the urban public can witness the mistral's dynamic materiality while also enjoying

Figure 6.1. In LUMA Arles's Parc des Ateliers, native trees and shrubs from surrounding biomes grow on small dunes that the mistral's gusts will continue to sculpt over time. Photo by author.

its cooling effects on the air (see figure 6.2). Around the pond, Smets has placed actual mirrors where park visitors can see themselves reflected in a microcosm of Provence's Mediterranean ecosystem. The park's trees—oaks, maples, and pines—are all native to the Mediterranean region, and its shrubs, grasses, and herbs were all selected from the park's three surrounding biotopes of the Alpilles, the Crau, and the Camargue.[3]

With its climate-attuned design, the Parc des Ateliers exemplifies a growing movement in contemporary Europe to establish new modes of coexistence between human beings and nature. In elevating the mistral to the position of co-creator of his public park, Smets and his team embraced a position of humble deference to local weather patterns and the ecosystems that they foster. From a biodiversity perspective, Smets's decision to integrate his park into its surrounding natural landscapes is already paying off. Wild birds en route to the Mediterranean coast have begun to appear on the park's grounds, pausing in its pond and its clusters of reeds. In its own small way, the park is helping to ensure that Provence continues its long-standing ecological role as a transit boulevard for the migratory creatures on their seasonal journey south.

*

The mistral-sculpted public park taking shape in contemporary Arles is a striking reminder that human attitudes toward the wind have always been contingent on shifting political, economic, and cultural priorities. Before the intensive modernization period of the nineteenth century, people in Provence had carefully adapted their livelihoods, dwellings,

Figure 6.2. The pond in the middle of the Parc des Ateliers serves numerous functions. It acts as a climate tool that cools the park in the summer heat while also attracting migrating birds with its beds of reeds. When it is stirred up by the mistral, the pond offers a place to witness the spectacular visual effects of gusting wind. Photo by author.

energy systems, and modes of transportation to the mistral's gusts. Thanks to their locally generated and embodied forms of climate knowledge, millers learned to crush their grain in sync with the wind, while sailors on the Rhône River figured out how to piggyback on the mistral's movement and zip downstream at great speed with their commercial goods in tow. On a deeper level, the mistral shaped traditional Provençal culture itself. The local sayings, myths, and rituals that gave meaning to everyday life in premodern Provence were closely intertwined with its howling presence.

In the period following the French Revolution, however, the national government in Paris took decisive actions to quell the mistral's forceful presence in the name of the public good. Through large-scale landscape modifications (afforestation and hedgerow planting), the introduction of new steam-powered technology for ships and mills, and the construction of a modern network of weather observatories, Provence became the locus of intensive experimentation in wind mitigation. The climate "improvements" that the French nation-state sought to achieve in Provence were not unique to that region; the modern environmental policies affecting the South of France were part of a much larger political and economic project to serve the national interest by subduing unruly regional environments and making the population less vulnerable to nature's whims.

Yet the central state's assertion of environmental control was never fully accepted in post-Revolutionary France, especially among provincial citizens. The nineteenth century witnessed a growing interest among regional medical, literary, and artistic communities in the healing and consoling power of nature. Whereas the French nation-building project called upon citizens to pledge their loyalty to an abstract set of universal political ideals, nineteenth-century French region builders promoted a grounded concept of territorial identity rooted in the authentic bodily experience of place. Rather than harming the public good and inhibiting civilizational progress, the mistral, from the vantage point of these regionalists, was essential to the well-being and communal identity of the Provençal people.

In contemporary Provence, the debate over the meanings and uses of the mistral continues to evolve. While civilian landscape architects like Bas Smets have begun to rethink the mistral's visibility in public spaces, the French government has also shifted its stance toward the wind. In late 2023, a floating wind farm called Provence-Grand-Large—comprised of three Spanish-designed Siemens Gamesa wind turbines—was completed off the coast of Marseille. Backed by France's state-owned energy company, Électricité de France (EDF), the wind farm, located seventeen kilometers from the shore and connected to a power station via cables installed beneath the floor of the Mediterranean Sea, is projected to generate 25 MW of energy for French consumers within the next decade.[4] Provence-Grand-Large reflects the French government's efforts to get on board with the European Union's goal of reaching 32 percent renewable energy consumption by 2030.[5] In a country known as the European leader in nuclear power, this floating wind farm signals the French state's new and significant openness to harnessing the renewable energy potential of France's diverse regional geography.

But even as the sleek new offshore turbines are beginning to generate electricity for Marseille, the future strength and geographic range of the wind that powers them remains unclear. Will the mistral reliably appear in coastal Provence to turn these futuristic turbines in the next ten, twenty, fifty, or one hundred years? According to recent scientific studies, the mistral, like several other local winds that blow through Europe, may be slowing down due to "anthropogenic forcing."[6] While the evidence is still emerging, the European Environmental Agency has been closely monitoring changes to wind speed and wind direction across the continent. The mistral's precarity thus joins other global-warming-related challenges that the French population will face in the coming decades, including hotter summers, raging wildfires, and severe droughts.[7]

In light of these current and future climate uncertainties, the French government recently called upon its regional councils—each of which represent one or more historic French provinces—to develop region-specific climate-adaptation plans for the next twenty to fifty years. Provence and several neighboring territories, which together form a larger administrative region called PACA (Provence-Alpes-Côte d'Azur), are currently formulating a multidecade climate-adaptation plan tailored to their geographic needs.[8] Heading this effort is the GREC-SUD (Groupe Régional d'Experts sur le Climat en Provence-Alpes-Côte D'Azur), a council of regional stakeholders that includes climate experts, university researchers, nonprofit associations, and citizens whose job is to help the southern region adapt its socioeconomic systems to a changing climate. Their initial reports have helped to identify the specific challenges facing regionally based industries such as wine and olive oil production, maritime ports, fisheries, and tourism.

In calling on leaders from a range of Provençal civil associations and public institutions to weigh in on climate-adaptation projects, the French government (acting within a European Union framework) has aligned itself with emerging scholarship on environmental resilience.

According to recent studies, people are far more likely to recognize the need for action on climate change when they can perceive its tangible effects on the localities where they live.[9] This is because climate plays an important role in defining a person's "comfort zone": their place of familiarity, childhood memories, internalized knowledge about the world, and physical sensibilities.[10] As this book has shown, distinct climate regions like windswept Provence have nurtured close physical and sentimental bonds between French civilians and their proximal environments since the nineteenth century. These close relationships remain visible today in public celebrations like the Festival of Wind, in Marseille, and in all the Provençal products—the perfumes, soaps, and corner cafés—that bear the name "mistral" with pride. To further climate-change awareness and action in contemporary Europe, it makes sense to strategically embrace the unofficial power of regional identities and to build on people's existing attachments to the familiar places and familiar winds that people know and love.

ACKNOWLEDGMENTS

France holds a special place of gratitude in my family history. During the 1920s, my relatives, refugees from the Russian Revolution, acquired Nansen passports and were offered residency in Paris, where they built new lives in the City of Light. My mother, born in 1943, became a French citizen during one of the nation's darkest hours. Though she boarded a steamer ship and moved to New York City with her family in the early 1950s, France left a long-lasting imprint on family traditions, including the bûche de Noël at Christmastime that became my job to bake. As a student, my curiosity for this faraway country that appeared in black-and-white family photographs led me to study abroad at the Stanford in Paris program during my junior year. Later, my personal map of France expanded far beyond the capital to the provinces, where the meaning of French identity stretches in a variety of directions as it mixes with different regional environments, histories, and cultures.

While working on this book, two formative experiences in Provence shaped my approach to writing. At the very beginning of the project, a residency at the Camargo Foundation in Cassis introduced me to a very special scholarly home nestled along the Mediterranean Sea. When I opened the door to my apartment, I couldn't believe the view. My balcony stood right above the Bay of Cassis and its glorious red-orange Cap Canaille. On windswept days, the balcony offered me an incredible front-row view of a mistral-blown sea. At Camargo, I was fortunate to receive the warm guidance of visionary director Julie Chénot and staff members Cécile Descloux and Julian McKinnon. Conversations with my talented cohort of fellows—James Cahill, Fiona McLaughlin, Amina Gautier, Timothy Brennan, Amanda Eurich, Aline Fischer, and Ilana Ortar—were essential in shaping the topics that the book would explore. Years later, bookending this project, I spent a very special week at the LUMA Foundation in Arles, at an environmental history conference organized by Grégory Quenet, Martin Guinard, and Maria Finders. The opportunity to visit with other environmental scholars after a long pandemic and to exchange ideas in the beautiful settings of Arles and the nearby Camargue wetlands was invaluable to getting this book across the finish line. It was also during this trip that Florence Sizaret and Anne-Marie Para, conservators at the newly reopened Museon Arlaten, in Arles, provided me with access to traditional wind-related objects (many of which appear in this book) that date back to Frédéric Mistral's original collections. I am also grateful for the assistance of staff members at the departmental archives of the Vaucluse in Avignon, the departmental archives of the Bouches-du-Rhône in Marseille, the municipal archives of Marseille, the National Archives in Paris, the National Library in Paris, the Alcazar Library in Marseille, and the Méjanes Library in Aix-en-Provence.

In addition to benefiting from the welcoming research environment that I found in France, this book was shaped by the fruitful six months that I spent as a writing fellow at the Rachel Carson Center in Munich. This remarkable place, led by Christof Mauch and Helmuth Trischler, offers a unique international setting in which to think collectively about how to best communicate environmental stories. My summer officemate at the RCC, Nancy Jacobs, has become a wonderful mentor. Sharing lunches outside in a nearby garden during a steamy Munich summer, visiting city sites with our families, and getting her early input on drafts was invaluable. Later in my stay, I benefited from support and creative input from Julia Tischler, Maryse Helbert, Anita Carrasco, Elena Feditchkina, Monica Vasile, Meredith McKittrick, Serenella Iovino, Roberta Biasillo, and Ruth Morgan. While in Germany, my husband and I also enjoyed the wonderful hospitality of Michael Layh, Katrin Layh, and their children. Thanks to them, we got to know the beautiful Allgäu region of Bavaria and to celebrate an authentic German Christmas.

Though it is a world away from Europe, Montana State University, in beautiful Bozeman, has helped to anchor the production of this book. A sabbatical in 2017–18 was essential for laying the groundwork for my archival research. In 2023, a Scholarship and Creativity Grant from the Vice President for Research Office offered a course buyout to help complete my final edits. I am grateful to my two department chairs over the past six years, Susan Cohen and Michael Reidy, for their continual support for funding this project—including an image subsidy for the book's color images—and for their moral support over wine and beer. In addition to my chairs, I have been lucky to be surrounded by many other friendly department colleagues, current and retired, who have offered helpful support and feedback over the years. Across the wider university, Jamie McEvoy has been a brilliant sounding board and a patient friend. Laura Burkle shared excellent meals and explained how wind fit into ecological webs. Adam Creitz and Jackson Rose at the Geospatial Core Facility at MSU generously drew the beautiful maps for the book. I would also like to thank my students at MSU, especially the undergraduates in my Nature & Culture in Europe course, for continuously rekindling my enthusiasm for this project with their fresh lines of thinking and thought-provoking questions.

Before I could finish this book, John Merriman, a mentor and a wonderful person, passed away. I told John about my idea for a mistral book during my last year of graduate school at Yale. I remember how he paused and considered the idea for a moment, and then enthusiastically encouraged me to do it. I miss John and his wife, Carol, who were as supportive and friendly as could be. I don't think any two Americans knew France better and cherished it more than they did. At crucial points in this book's journey, conversations with French historians Kieko Matteson, Fabien Locher, Peter McPhee, Bruno Cabanes, Caroline Ford, Andrea Duffy, and Sarah Griswold were helpful for grasping the mistral's role in shaping the past. In the environmental history community, I've benefited from the kindness and mentorship of Jeremy Vetter, Kathy Brosnan, Jim Akerman, Finis Dunaway, Mark Hersey, and Stephen Brain. Friends from the Bay Area, Justin Virgili and KC George and their families, and Bríd Arthur, Sonali Bhagat, Keira Gipson, and Zoë Chafe have been there, even if remotely, to offer good cheer. Many thanks go to my acquisitions editor, Susannah Engstrom, my production editor, Stephen Twilley, and to the entire staff at the University of Chicago Press for helping to shepherd this book to its completion.

Finally, I would like to thank my family for their unwavering support. My beloved father, John Dunlop, passed away shortly before I received the copy edits for this book. One of his favorite things to do, as a retired professor, was to clip newspaper articles related to France and the environment to help further my research in any way he could. Together with my mother, Olga Dunlop, he encouraged me in my path as a writer and a scholar from the very beginning. May his memory be eternal. My mother, who continues to provide invaluable support, has been waiting for this book for quite some time, and I hope she is pleased with the result. My siblings Maria, John, and Olga, their spouses Peter, Bea, and Nick, as well as my nieces and nephews John, Lizzie, Sophia, Gregory, Jacob, Olga, and Serge, have been wonderful company, as have my Johnson in-laws who welcomed me warmly into their family. The joyful morning wags of Penny, our one-eyed blue heeler from somewhere in Montana, helped me to keep moving forward each day. Most of all, I wish to thank my husband, Erick Johnson, for his steadfast support. Years ago, a fortuitous meeting in his office helped me to begin thinking about the mistral as a powerful fluid that flowed across Provence. As I researched and wrote this book, he was my vital sounding board. Like a sea otter, he held on strong, making sure I didn't drift away into the abyss of the writing process. This was not the easiest period to complete a book, but his empathy, counsel, wise input, and cooking made all the difference. This book is dedicated to him.

NOTES

Introduction

1. Émile Zola, *Doctor Pascal*, trans. Julie Rose (Oxford: Oxford University Press, 2020), 224.

2. Charles Martins, *Le Mont-Ventoux en Provence* (Paris: Imprimerie de J. Claye, 1863), 6. Unless otherwise noted, all translations are my own.

3. Alphonse Daudet, *Tartarin of Tarascon* (Boston: T. Y. Crowell, 1895), 92.

4. The map of the mistral windscape (figure 0.1) and the map of Provence (figure 0.2) were created using ArcGIS software by Esri. ArcGIS and ArcMap are the intellectual property of Esri and are used herein under license. Copyright © Esri. All rights reserved. For more information about Esri software, please visit www.esri.com. The rivers featured on the maps were created using OpenStreetMap, which is open data and licensed under the Open Data Commons Open Database License (ODbL) by the OpenStreetMap Foundation (OSMF).

5. David Abulafia, *The Great Sea: A Human History of the Mediterranean* (Oxford: Oxford University Press, 2011), xxviii.

6. Victor Gelu, *Oeuvres complètes de Victor Gelu*, vol. 1 (Paris: G. Charpentier, 1886), lv.

7. Horace Bénédict de Saussure, *Voyages dans les Alpes* (Neuchâtel: Louis Fauche-Borel, 1796), 404.

8. Vincent van Gogh to Émile Bernard, in *Vincent van Gogh: The Letters*, ed. Leo Jansen, Hans Luijten, and Nienke Bakker, letter 698, accessed December 14, 2020, www.vangoghletters.org/vg/letters/let698/letter.html.

9. For a stunning photographic album of scenes from mistral-blown Provence, see Rachel Cobb, *Mistral: The Legendary Wind of Provence* (Bologna: Damiani, 2018).

10. Jean-Luc Massot, *Maisons rurales et vie paysanne en Provence* (Paris: Serg-Berger-Levrault, 1992), 30.

11. A mistral event occurs if wind gusts surpass 35.8 mph (57.6 kmph) from a directional orientation between 320 and 030 degrees. See Valérie Jacq, Philippe Albert, and Robert Delorme, "Le mistral: Quelques aspects des connaissances actuelles," *La Météorologie* 50 (August 2005): 35–36.

12. Jules Michelet, *History of France from the Earliest Period to the Present Time*, vol. 1, trans. G. H. Smith (New York: D. Appleton, 1851), 165.

13. Joseph Conrad, *The Arrow of Gold: A Story between Two Notes* (New York: Doubleday, 1921), 65.

14. One notable exception is Alain Corbin's scholarship on the history of weather, emotion, and the senses in modern France. For his most recent book, see *La Rafale et le zéphyr: Histoire des manières d'éprouver et de rêver le vent* (Paris: Fayard, 2021), 29–30. See also Fernand Braudel's brief discussion of the mistral in his longue durée history of the Mediterranean world, originally published in 1949, *The Mediterranean and the Mediterranean World in the Age of Philip II*, vol. 1, trans. Siân Reynolds (Berkeley: University of California Press, 1995), 250–53. Several local histories of the mistral geared for regional French audiences have also been published. See Bernard Mondon and Steffen Lipp, *Petite anthologie du mistral* (Saint-Rémy-de-Provence: Éditions Équinoxe, 2004); Charles Galtier, *Météorologie populaire dans la France ancienne: La Provence, Empire du soleil et royaume des vents* (Le Coteau: Éditions Horvath, 1984); Maurice Pezet, *La Provence sous le Mistral* (Avignon: Éditions les Chants du Rhône, 1983); and Emmanuel Davin, *Mistral et autres vents en Provence* (Toulon: Société nouvelle des imprimeries toulonnaises, 1938).

15. My use of the term "silence" draws from J. B. Harley's argument that maps routinely silence or erase aspects of a landscape while emphasizing other parts. See "Silences and Secrecy: The Hidden Agenda of Cartography in Early Modern Europe," chap. 3 in *The New Nature of Maps: Essays in the History of Cartography* (Baltimore: Johns Hopkins University Press, 2001).

16. The term "windscape" appears frequently in scientific literature on place-based animal behavior—including flight patterns in birds and foraging behavior in large carnivores—as well as in studies on wind power potential in specific regions. See Françoise Amélineau et al., "Windscape and Tortuosity Shape the Flight Costs of Northern Gannets," *Journal of Experimental Biology* 217, no. 6 (March 2014): 876–85; Ron R. Togunov, Andrew E. Derocher, and Nicholas J. Lunn, "Windscapes and Olfactory Foraging in a Large Carnivore," *Scientific Reports* 7, no. 46332 (2017), https://doi.org/10.1038/srep46332; Giovanni Mauro, "The New 'Windscapes' in the Time of Energy Transition: A Comparison of Ten European Countries," *Applied Geography* 109, no. 102041 (2019), https://doi.org/10.1016/j.apgeog.2019.102041.

17. The mistral's close ties to the terrain of southern France put it into the same category as other small-scale winds that are particular to mountain geographies in diverse parts of the globe, including the foehn of Austria and Switzerland, the Santa Ana of Southern California, the khamsin of North Africa, and the Helm wind of northern England. These "local winds" share not only the imprint of regional geography but also a capacity to structure social relationships, religious beliefs, and settlement patterns in the areas in which they blow. For a comparative discussion of local European winds, see Nick Hunt, *Where the Wild Winds Are: Walking Europe's Winds from the Pennines to Provence* (Boston: Nicholas Brealey Publishing, 2017), and, more generally, Lyall Watson, *Heaven's Breath: A Natural History of the Wind* (New York: New York Review Books, 2019). On specific local winds, see Sarah Strauss, "An Ill Wind: The Foehn in Leukerbad and Beyond," *Journal of the Royal Anthropological Institute* 13 (2007): 165–81; Lucy Veale, Georgina Endfield, and Simon Naylor, "Knowing Weather in Place: The Helm Wind of Cross Fell," *Journal of Historical Geography* 45 (2014): 25–37; and, for an evocative description of the Santa Ana wind's influence on daily life in 1960s Los Angeles, Joan Didion's "Los Angeles Notebook," in *Slouching Towards Bethlehem* (New York: Farrar, Straus and Giroux, 1968).

18. Wind direction refers to the location from which the wind is blowing. The mistral emerges from the northwest and moves downward to the Mediterranean Sea.

19. Fernand Braudel, *The Identity of France*, vol. 1, trans. Siân Reynolds (New York: Harper and Row, 1988), 31.

20. Braudel, *Identity of France*, 32.

21. On the geomorphology of the Mediterranean landscape and its relationship to climate, see Harriet D. Allen, *Mediterranean Ecogeography* (London: Pearson, 2001), 38; Jacques Blondel and James Aronson, *Biology and Wildlife in the Mediterranean Region* (Oxford: Oxford University Press, 1999), 2.

22. *Géographie de Strabon: Traduite de grec en français*, vol. 2 (Paris: Imprimerie Impériale, 1809), 18.

23. Michel Darluc, *Histoire naturelle de Provence: Contenant ce qu'il y a de plus remarquable dans les règnes Végétal, Minéral, Animal & la partie Géoponique* (Avignon: J. J. Niel, 1792), 296.

24. "Du Mistral (Son ancienneté): Explication du phénomène," in *Mémoires de l'Académie de Vaucluse* 9 (1890): 305.

25. *Oxford English Dictionary Online*, s.v. "mistral (*n.*)," revised 2002, https://www.oed.com/dictionary/mistral_n.

26. Tim Ingold uses the term "weather world" to convey the place-making power of localized weather conditions. See "Footprints through the Weather World: Walking, Breathing, Knowing," *Journal of the Royal Anthropological Institute* 16, no. 1 (May 2010): 121–39.

27. Environmental historians have used the concept of "unruliness" to explore the challenges that modern states face in establishing control over physical environments. See Siddhartha Krishnan, Christopher L. Pastore, and Samuel Temple, "Introduction: Unruly Matters," in "Unruly Environments," ed. Siddhartha Krishnan, Christopher L. Pastore, and Samuel Temple, special issue, *Rachel Carson Center Perspectives*, no. 3 (2015): 5–7.

28. On the transformation of France's territorial administration during the Revolution of 1789, see Marie-Vic Ozouf-Marignier, *La formation des départements: La représentation du territoire français à la fin du 18e siècle* (Paris: Éditions de l'École des hautes études en sciences sociales, 1989).

29. On the relationship between French identity and territory, see Paul Claval, "From Michelet to Braudel: Personality, Identity and Organization of France," in *Geography and National Identity*, ed. David Hooson (Oxford: Blackwell, 1994).

30. On the resilience of ecoregions in nineteenth-century France, see Daniel A. Finch-Race and Valentina Gosetti, "Discovering Industrial-Era French Ecoregions," *Dix-Neuf* 23, nos. 3–4 (2019): 151–62.

31. On the sand dunes of the Landes, see Diana K. Davis, *Arid Lands: History, Power, Knowledge* (Cambridge, MA: MIT Press, 2016), 76–78; Caroline Ford, *Natural Interests: The Contest over Environment in Modern France* (Cambridge, MA: Harvard University Press, 2016), 63–64.

32. See Julian Jackson, *Paris under Water: How the City of Light Survived the Great Flood of 1910* (New York: Palgrave MacMillan, 2010).

33. James C. Scott argues that at the center of the high modern state's ideology was "a supreme self-confidence about continued linear progress, the development of scientific and technical knowledge, the expansion of production, the rational design of social order, the growing satisfaction of human needs, and . . . an increasing control over nature." *Seeing Like a State: How Certain Schemes to Improve the Human Condition Have Failed* (New Haven, CT: Yale University Press, 1998), 90. For a more recent critique of the risks involved with the modern state's increased technological control over environments, see Jean-Baptiste Fressoz, *L'Apocalypse joyeuse: Une histoire du risque technologique* (Paris: Seuil, 2012).

34. On regional and local-scale environmental and climate studies in nineteenth-century France, see Stéphane Gerson, *The Pride of Place: Local Memories and Political*

Culture in Nineteenth-Century France (Ithaca, NY: Cornell University Press, 2003), 83. See also Jean-Baptiste Fressoz and Fabien Locher, *Les Révoltes du ciel: Une histoire du changement climatique XVe–XXe siècle* (Paris: Seuil, 2020), 141–58.

35. On my use of outdoor archives to research the history of human-mistral interactions, see Catherine T. Dunlop, "Losing an Archive: Doing Place-Based History in the Age of the Anthropocene," *American Historical Review* 126, no. 3 (September 2021): 1143–53.

36. Afforestation became a popular nineteenth-century "solution" for slowing winds. On the intersection of activist forestry and political culture in modern France, see Kieko Matteson, *Forests in Revolutionary France: Conservation, Community, and Conflict, 1669–1848* (Cambridge: Cambridge University Press, 2015); Andrea E. Duffy, *Nomad's Land: Pastoralism and French Environmental Policy in the Nineteenth-Century Mediterranean World* (Lincoln: University of Nebraska Press, 2019).

37. On the economic and technological consolidation of French national space during the mid-nineteenth century, see David Harvey, *Paris: Capital of Modernity* (New York: Routledge, 2005).

38. The fact that natural forces do not have intention or desire is irrelevant—the physical impact they have is sufficient to include them as historical agents. See recent neomaterialist scholarship, including Jane Bennett, *Vibrant Matter: A Political Ecology of Things* (Durham, NC: Duke University Press, 2010); Timothy J. LeCain, *The Matter of History: How Things Create the Past* (Cambridge: Cambridge University Press, 2017). For a discussion of nature's "agency" in history, see Linda Nash, "The Agency of Nature or the Nature of Agency?" *Environmental History* 10, no. 1 (January 2005): 67–69.

39. Élisée Reclus, *The Ocean, Atmosphere, and Life*, trans. B. B. Woodward (New York: Harper and Bros., 1874), 179.

40. For an excellent study on the development of regional identity in Provence, see Robert Zaretsky, *Cock and Bull Stories: Folco de Baroncelli and the Invention of the Camargue* (Lincoln: University of Nebraska Press, 2004).

41. For a recent study on the growth of "environmental awareness" in modern France, see Ford, *Natural Interests*.

42. See, for example, Brian C. J. Singer, "Cultural versus Contractual Nations: Rethinking Their Opposition," *History and Theory* 35, no. 3 (October 1996): 309–37.

43. Ernest Renan, *"Qu'est-ce qu'une nation?": Conférence faite en Sorbonne, le 11 mars 1882* (Paris: Calmann Lévy, 1882).

44. Benedict Anderson, *Imagined Communities: Reflections on the Origin and Spread of Nationalism* (London: Verso, 2016).

45. On the emerging distinctions between the *grande patrie* (big homeland) and the *petite patrie* (little homeland) in modern France, see Anne-Marie Thiesse, *Ils apprenaient la France: L'Exaltation des régions dans le discours patriotique* (Paris: Éditions de la Maison des Sciences de l'Homme, 1997); Jean-François Chanet, *L'École républicaine et les petites patries* (Paris: Aubier, 1996); Gerson, *Pride of Place*; Catherine T. Dunlop, *Cartophilia: Maps and the Search for Identity in the French-German Borderland* (Chicago: University of Chicago Press, 2015).

46. Jessica Hayes-Conroy and Allison Hayes-Conroy, "Visceral Geographies: Mattering, Relating, and Defying," *Geography Compass* 4, no. 9 (2010): 1273–83.

47. See Ingold, "Footprints through the Weather World."

48. See Helen Lefkowitz Horowitz, *A Taste for Provence* (Chicago: University of Chicago Press, 2016). For an excellent critical take on postwar tourism in Provence, see

Stephen L. Harp, *The Riviera, Exposed: An Ecohistory of Postwar Tourism and North African Labor* (Ithaca, NY: Cornell University Press, 2022).

49. I interpret the mistral's impact on human beings not in terms of determining how they acted, but as an atmospheric presence that shaped their possible political, economic, and cultural choices. Paul Vidal de la Blache, the leading nineteenth-century geographer of France's Third Republic, used the term "possibilism" to describe the choices that people have always had in adapting to their local environmental conditions. He helped to develop the notion of possibilism to counteract dangerous notions of environmental determinism emerging in Germany during the same turn-of-the-century period through the work of Friedrich Ratzel, among others. For a discussion of the problems with deterministic thinking in geography and political ecology, see Paul Robbins, *Political Ecology: A Critical Introduction* (Hoboken, NJ: Wiley-Blackwell, 2020), 23–25. For an example of Vidal de la Blache's possibilist approach to geography, see Paul Vidal de la Blache, *Principles of Human Geography*, trans. Milicent Todd Bingham (New York: Henry Holt, 1926).

50. Chapter 5 of this book is adapted from my article "Looking at the Wind: Paintings of the Mistral in Fin-de-Siècle France," *Environmental History* 20, no. 3 (2015): 505–18. © 2015 by Catherine T. Dunlop.

Chapter One

1. Alexandre Dumas, *Pictures of Travel in the South of France* (London: Offices of the National Illustrated Library, 1851), 141.

2. Proverbs cited in Jean-Philippe Chassany, *Dictionnaire de météorologie populaire* (Paris: Maisonneuve et Larose, 1989), 391.

3. European Environmental Agency, "Biogeographical Regions in Europe," last modified October 30, 2017, https://www.eea.europa.eu/data-and-maps/figures/biogeographical-regions-in-europe-2.

4. For a discussion of the origins and behavior of winds in the Mediterranean ecosystem, see Allen, *Mediterranean Ecogeography*, 38; Jacques Blondel and James Aronson, *Biology and Wildlife in the Mediterranean Region* (Oxford: Oxford University Press, 1999), 2.

5. Blondel and Aronson, *Biology and Wildlife*, 21.

6. Daniel W. Gade, "Windbreaks in the Lower Rhône Valley," *Geographical Review* 68, no. 2 (April 1978): 132.

7. For a detailed discussion of plant life in Provence's coastal Calanques region, see Anne Merry and Bruno Lambert, *Flore des Calanques* (Les Sablonnières: Éditions du Fournel, 2014).

8. See A. T. Grove and Oliver Rackham, *The Nature of Mediterranean Europe: An Ecological History* (New Haven, CT: Yale University Press, 2003), 72–117.

9. Allen, *Mediterranean Ecogeography*, 10–11.

10. On the concept of hybrid landscapes, see Mark Fiege, *Irrigated Eden: The Making of an Agricultural Landscape in the American West* (Seattle: University of Washington Press, 1999), 9.

11. For an excellent study on "working landscapes" in Europe, see Sarah R. Hamilton, *Cultivating Nature: The Conservation of a Valencian Working Landscape* (Seattle: University of Washington Press, 2018).

12. For a cross-cultural account of peasant life in the nineteenth-century Mediterranean Basin, see Duffy, *Nomad's Land.*

13. Richard White, "'Are You an Environmentalist or Do You Work for a Living?': Work and Nature," in *Uncommon Ground: Rethinking the Human Place in Nature*, ed. William Cronon (New York: W. W. Norton, 1996), 171–85.

14. Robert Darnton, *The Great Cat Massacre and other Episodes in French Cultural History* (New York: Vintage Books, 1985), 5.

15. Definitions cited in Frédéric Mistral, *Lou Tresor dóu Felibrige ou Dictionnaire provençal-français*, vol. 2 (Aix-en-Provence: Imprimerie Veuve Remondet-Aubin, 1878), 347.

16. Proverbs cited in Chassany, *Dictionnaire de météorologie populaire*, 231.

17. The archive's finding aid indicates that the journal was once owned by a farmer from the district of Arles named Astier. See "Agenda du commerce, de l'industrie et des besoins journaliers," 1890, Archives départementales des Bouches-du-Rhône, 1 J 838.

18. See Thomas Carter and Elizabeth Collins Cromley, *Invitation to Vernacular Architecture: A Guide to the Study of Ordinary Buildings and Landscapes* (Knoxville: University of Tennessee Press, 2005), xviii–xx. I would also like to thank Dr. Janet Ore for her helpful comments on the intersection of vernacular architecture and environmental history.

19. Ian McHarg, *Design with Nature* (New York: Natural History Press, 1969), 5.

20. Fernand Benoit, *La Provence et le Comtat Venaissin: Arts et traditions populaires* (Avignon: Aubanel, 1975), 44.

21. See Jean Mercier, "L'habitation rurale provençale: Le vent et le soleil; Quelques remarques préliminaires," *Revue de géographie alpine* 31, no. 4 (1943): 525–33.

22. Benoit, *La Provence et le Comtat Venaissin*, 62.

23. Daudet, *Letters from My Mill*, 176.

24. Tim Ingold, *The Perception of the Environment: Essays on Livelihood, Dwelling, and Skill* (London: Routledge, 2011), 175.

25. Benoit, *La Provence et le Comtat Venaissin*, 77–78.

26. Daudet, *Letters from My Mill*, 176–77.

27. Jean-Luc Massot, *Maisons rurales et vie paysanne en Provence* (Paris: Serg-Berger-Levrault, 1992), 81.

28. On nomadic shepherds in nineteenth-century Provence, see Duffy, *Nomad's Land.*

29. Benoit, *La Provence et le Comtat Venaissin*, 321–22.

30. On the history of seasonal movements from mountains to plains, see John McNeill, *The Mountains of the Mediterranean World: An Environmental History* (Cambridge: Cambridge University Press, 2003). See also Duffy, *Nomad's Land.*

31. Massot, *Maisons rurales et vie paysanne en Provence*, 63.

32. Benoit, *La Provence et le Comtat Venaissin*, 164.

33. Steven Laurence Kaplan, *Provisioning Paris: Merchants and Millers in the Grain and Flour Trade during the Eighteenth Century* (Ithaca, NY: Cornell University Press, 1984), 222.

34. Kaplan, *Provisioning Paris*, 222.

35. Kaplan, 222.

36. Jean-Marie Homet, *Provence des moulins à vent* (Aix: Edisud, 1984), 19.

37. Alphonse Jouven, *Sur les moulins à farine* (Aix: Imprimerie de Nicot et Pardigon, 1848), 209.

38. Kaplan, *Provisioning Paris*, 244.

39. Kaplan, 247.

40. The figure of 22,137,500 trees reflects the plants coming from state nurseries or bought by the administration from private nurseries. Ministère des Forêts, *Reboisement*

des montagnes (loi du 29 juillet 1860): Compte rendu des travaux de 1862 (Paris: Imprimerie Impériale, 1863), 23.

41. Peter M. Jones, *Agricultural Enlightenment: Knowledge, Technology, and Nature, 1750–1840* (Oxford: Oxford University Press, 2016), 6.

42. See Diana K. Davis, "Exploration, Desiccation, and Improvement," in *The Arid Lands: History, Power, Knowledge* (Cambridge, MA: MIT Press, 2016), 49–79; Grove and Rackham, "Introduction: Ruined Landscapes and the Question of Desertification," in *The Nature of Mediterranean Europe*, 8–24.

43. For an in-depth discussion of Rauch's scientific influence, see Ford, "François-Antoine Rauch's New Harmony of Nature," in *Natural Interests*, 16–42; Davis, *The Arid Lands*, 65–67.

44. F. A. Rauch, *Harmonie Hydro-végétale et météorologique* (Paris: Frères Levrault, 1802), 110.

45. Rauch, *Harmonie Hydro-végétale et météorologique*, 110.

46. On the French state's involvement in climate research during the early nineteenth century, see Fabien Locher and Jean-Baptiste Fressoz, "Modernity's Frail Climate: A Climate History of Environmental Reflexivity," *Critical Inquiry* 38, no. 33 (Spring 2012): 579–80. See also Emmanuel Le Roy Ladurie, *Times of Feast, Times of Famine: A History of Climate since the Year 1000*, trans. Barbara Bray (Garden City, NY: Doubleday, 1971), 3.

47. For a discussion of this state-sponsored French climate change survey, see Jean-Baptiste Fressoz and Fabien Locher, *Les Révoltes du ciel: Une histoire du changement climatique XVe–XXe siècle* (Paris: Seuil, 2020), 141–58.

48. Letter from the Mairie de Marseille, Division des Archives, to the Préfet, no. 115, "Renseignements sur le déboisement des montagnes," Marseille, June 7, 1821, Archives départementales des Bouches-du-Rhône, 7 M 163.

49. Mairie de Marseille to the Préfet, no. 115.

50. Letter from the Sous-Préfecture d'Aix, Bouches-du-Rhône, Bureau de Commerce, to the Préfet du Département des Bouches-du-Rhône, May 17, 1822, Archives départementales des Bouches-du-Rhône, 7 M 163.

51. Letter from Arles, Bureau de Commerce, to the Préfet du Département des Bouches-du-Rhône, April 22, 1822, Archives départementales des Bouches-du-Rhône, 7 M 163.

52. Letter from the Mairie d'Aubagne to the Préfet du Département des Bouches-du-Rhône, Comte de Villeneuve, May 22, 1821, Archives départementales des Bouches-du-Rhône, 7 M 163.

53. According to Kieko Matteson, France's wooded domain was one hundred thousand hectares smaller in 1800 than it had been in 1789. "Peasant pilfering, incursion, and abuse played a role in this decline, but more significant were the auctions of nationalized woodland properties to buyers—principally better-off landowners, manufacturers, and bourgeois—who quickly put them to the axe." *Forests in Revolutionary France: Conservation, Community, and Conflict, 1669–1848* (Oxford: Oxford University Press, 2015), 152–53.

54. Letter from the Mairie de Marseille to the Préfet du Département des Bouches-du-Rhône, Marseille, March 26, 1821, Archives départementales des Bouches-du-Rhône, 7 M 163; André Thoüin, *Instruction sur l'établissement des pépinières, leur distribution, leur culture et leur usage* (Paris, 1822), Archives départementales des Bouches-du-Rhône, 7 M 163; For a summary, see L.-G. Delamarre, *Traité pratique de la culture des pins* (Paris: Chez Madame Huzard, 1831), 14–38. On the development of French forestry techniques in the nineteenth century, see Duffy, *Nomad's Land*.

55. Prosper Demontzey, *Traité pratique du reboisement et du gazonnement des montagnes* (Paris: J. Rothschild, 1882), 141.

56. “Rapport relative aux semis de grains de Pin Laricio,” Recueil administrative du département des Bouches-du-Rhône, no. 29, 1821, Archives départementales des Bouches-du-Rhône, 7 M 163.

57. Letter from the Administration des Forêts to the Préfet du Département des Bouches-du-Rhône, August 21, 1847, Archives départementales des Bouches-du-Rhône, 7 M 163.

58. Letter from the Administration des Forêts to the Préfet du Département des Bouches-du-Rhône, August 21, 1847.

59. Frédéric Mistral, *Mes origines: Mémoires et récits* (Arles: Actes Sud, 2008), 23.

60. Daniel W. Gade, “Windbreaks in the Lower Rhône Valley,” *Geographical Review* 68, no. 7 (April 1978): 133.

61. Statistic cited in Louis Castagne, *Observations sur le reboisement des montagnes et des terrains vagues, dans le département des Bouches-du-Rhône* (Aix: Vve Taverin, 1851), 480.

62. Gade, “Windbreaks in the Lower Rhone Valley,” 134.

63. Antoine-César Becquerel, *Des climats et de l'influence qu'exercent les sols boisés et non boisés* (Paris: Didot Frères, 1853), 116.

64. George Perkins Marsh, quoted in Gade, “Windbreaks in the Lower Rhone Valley,” 144.

65. Henri Laure, *Manuel du cultivateur provençal ou Cours d'agriculture simplifiée pour le midi de l'Europe et le Nord de l'Afrique* (Toulon: Chez Teisseire, 1839), 673.

66. Editor's response to letter in the *Journal d'agriculture pratique* (July 1895), 157.

67. L. Reich, “Camargue,” *Journal d'agriculture pratique* (January 1877), 124.

68. Paul Granger, *Les Fleurs du Midi: Cultures florales industrielles* (Paris: Librairie J.-B. Baillière et fils, 1902), 3.

69. See, for example, Zaretsky, *Cock and Bull Stories*.

70. Cited in Michel de Certeau, Dominique Julia, and Jacques Revel, *Une politique de la langue: La Révolution française et les patois; L'enquête de Grégoire* (Paris: Gallimard, 1975), 142–43.

71. Certeau, Julia, and Revel, *Une politique de la langue*, 148.

72. Certeau, Julia, and Revel, 14.

73. Certeau, Julia, and Revel, 136.

74. Matteson, *Forests in Revolutionary France*, 66.

75. On preservationist movements in nineteenth-century France, see Ford, *Natural Interests*, 15.

76. See Deborah Silverman, *Van Gogh and Gauguin: The Search for Sacred Art* (New York: Farrar, Straus and Giroux, 2000), 53.

77. Poem to J. A. Vaisse, March 24, 1845, cited in *La Provence rurale de 1850 à 1900 vue par ses peintres, ses félibres, et ses poètes*, ed. Paule Brahic and Pierre Echinard (Marseille: Fondation regards de Provence, 2005).

78. Frédéric Mistral, *Mirèio: Pouèmo prouvençau / Mireille: Poème provençal* (Paris: Charpentier Libraire-Éditeur, 1868), 281.

79. Mistral, *Mirèio/Mireille*, 331.

80. Mistral, *Mes origines*, 9.

81. “Old Cornille's Secret,” in Alphonse Daudet, *Letters from My Windmill*, trans. Dédicace (Rennes: Éditions Ouest-France, 2011), 7.

82. Daudet, *Letters from My Windmill*, 8.

83. Daudet, 8.

84. Models of all these structures appeared at the Museon Arlaten, founded in 1899 by Frédéric Mistral and Émile Marignan. See Véronique Dassié and Dominique Séréna-Allier, "Are Popular Local Artefacts Exotic? Building 'Provençalness' at the Museon Arletan [*sic*]," *Journal of Museum Ethnography*, no. 22 (December 2009): 129–41.

Chapter Two

1. Émile Zola, *Naïs Micoulin* (Paris: G. Charpentier, 1883), 40.

2. Regarding the mistral's impact on the thermal conditions of the Mediterranean, see Jacques Collina-Girard, "Underwater Landscapes and Implicit Geology, Marseilles and the National Calanques Park," in *Underwater Seascapes: From Geological to Ecological Perspectives*, ed. Olivier Musard et al. (London: Springer, 2014), 68.

3. Fernand Braudel, *The Mediterranean and the Mediterranean World in the Age of Philip II*, vol. 1, trans. Siân Reynolds (Berkeley: University of California Press, 1996), 233.

4. Abulafia, *Great Sea*, xxviii.

5. Gunther Peck uses the concept of "geography of labor" to explore the "spatial, material, and cultural connections between nature and labor." See "The Nature of Labor: Fault Lines and Common Ground in Environmental and Labor History," *Environmental History* 11, no. 2 (April 2006): 214.

6. On steam transportation's impact on travel times between Mediterranean ports, see Julia Clancy-Smith, *Mediterraneans: North Africa and Europe in an Age of Migration, c. 1800–1900* (Berkeley: University of California Press, 2011), 27.

7. Alain Corbin, *The Lure of the Sea: The Discovery of the Seaside in the Western World, 1750–1840*, trans. Jocelyn Phelps (New York: Penguin, 1994), 8–9.

8. Fernand Braudel, *Memory and the Mediterranean*, trans. Siân Reynolds (New York: Vintage, 2002), 13.

9. On the cultural construction of geography, see Martin W. Lewis and Kären E. Wigen, *The Myth of Continents: A Critique of Metageography* (Berkeley: University of California Press, 1997).

10. See *Mémoire des rives: Cartes et portulans de Méditerranée* (Marseille: Bibliothèque Alcazar, 2013).

11. On the tensions between Marseille merchants and the Bourbon monarchy, see Junko Thérèse Takeda, *Between Crown and Commerce: Marseille and the Early Modern Mediterranean* (Baltimore: Johns Hopkins University Press, 2011).

12. John H. Pryor, *Geography, Technology and War: Studies in the Maritime History of the Mediterranean, 649–1571* (Cambridge: Cambridge University Press, 1992), 34.

13. Pryor, *Geography, Technology and War*, 53

14. Patricia Payn-Echalier and Philippe Rigaud, *Pierre Giot: Un capitaine marin arlésien 'dans la tourmente'; Journal, livre de bord, correspondance, 1792–1816* (Aix-en-Provence: Presses universitaires de Provence, 2016), 41.

15. Payn-Echalier and Rigaud, *Pierre Giot*, 41.

16. Braudel, *The Mediterranean*, 106.

17. Braudel, 90.

18. Statistic cited in William H. Sewell Jr., *Structure and Mobility: The Men and Women of Marseille, 1820–1870* (Cambridge: Cambridge University Press, 2009), 22.

19. François-Frédéric Leméthéyer, *Dictionnaire moderne des termes de marine et de la navigation à vapeur* (Paris: Robiquet, 1843), 326

20. Payn-Echalier and Rigaud, *Pierre Giot*, 145.

21. For a description of the types of environmental challenges involved in sailing between Arles and Marseille, including winds, see the letter from ship captains and merchants in Arles to the president of the chamber of commerce in Marseille, October 10, 1836, Archives de la Chambre de Commerce de Marseille, M.R. 4.

22. Ship captains and merchants in Arles to the president of the chamber of commerce in Marseille, October 10, 1836, 29.

23. For a description of *devino-vènts*, see J. Bourrilly and Louis Aubert, *Objets et rites talismaniques en Provence d'après les collections du Museon Arlaten* (Valence: Imprimerie Valentinoise, 1907).

24. Henri Michelot, *Portulan de la mer méditerranée ou guide des pilotes côtiers* (Marseille: Jean Mossy, 1805), iv.

25. Michelot, *Portulan de la mer méditerranée*, xii.

26. Karl Marx, *Capital*, trans. Samuel Moore and Edward Aveling, vol. 1 (New York: International Publishers, 1992), 173.

27. Marx, *Capital*, 174.

28. Corbin, *Lure of the Sea*, 209.

29. Sabin Berthelot, *Étude sur les pêches maritimes dans la méditerranée et l'océan* (Paris: Challamel Ainé, 1868), 7.

30. For an account of Antoine Roux's life and artistic work, see Jean Meissonier, *Voiliers de l'époque romantique peints par Antoine Roux* (Lausanne: Edita Lausanne, 1968); Philip Chadwick Foster Smith, *The Artful Roux: Marine Painters of Marseille* (Salem, MA: The Peabody Museum of Salem, 1978).

31. E. A. Wrigley, *The Path to Sustained Growth: England's Transition from an Organic Economy to an Industrial Revolution* (Cambridge: Cambridge University Press, 2016), 2.

32. Wrigley, *Path to Sustained Growth*, 134.

33. Julia A. Clancy-Smith, *Mediterraneans: North Africa and Europe in an Age of Migration, c. 1800–1900* (Berkeley: University of California Press, 2011), 25.

34. Cited in Clancy-Smith, *Mediterraneans*, 27.

35. By the 1830s, Marseille was the fifth-largest port in the world, trailing only New York City, London, Liverpool, and Hamburg in the quantity of cargoes unloaded onto its docks. Cited in Sewell, *Structure and Mobility*, 19.

36. Jean-Jacques Antier, *Marins de Provence et du Languedoc: Vingt-cinq siècles d'histoire du littoral français méditerranéen* (Montpellier: Les Presses de Languedoc, 2003), 234.

37. Michel Chevalier, *Essais de politique industrielle* (Paris: Librairie de Charles Gosselin, 1843), 144–45.

38. Achille-François-Léonor de Jouffroy d'Abbans, *Des Bateaux à vapeur: Précis historique de leur invention* (Paris: Imprimerie de E. Duvenger, 1839), 49.

39. A. Campaignac, *De l'état actuel de la navigation par la vapeur* (Paris: Librairie industrielle-scientifique, 1842), 13.

40. Galy Cazalat, *Mémoire théorique et pratique sur les bateaux à vapeur* (Paris: Librairie scientifique et industrielle de L. Mathias, 1837), 162.

41. Jules Julliany, *Essai sur le commerce de Marseille* (Marseille: Jules Barile, 1842), 246.

42. Earlier, when the French Crown ordered the production of a massive arsenal of galley ships in seventeenth- and eighteenth-century Marseille, the city built several immense rope-making factories (*corderies*) in addition to a sail-making factory (*voilerie*), which produced wool and cotton sails for royal military ships. See Antier, *Marins de Provence et du Languedoc*, 17.

43. Sewell, *Structure and Mobility*, 49–50.

44. Payn-Echalier and Rigaud, *Pierre Giot*, 50.

45. On the rising fear of accidents due to steam transportation, see Wolfgang Schivelbusch, *The Railway Journey: The Industrialization of Time and Space in the Nineteenth Century* (Berkeley: University of California Press, 2014).

46. See examples of navigation permits, or *permis de navigation*, in "Commission de surveillance des bateaux à vapeur à Marseille et Arles," Archives départementales des Bouches-du-Rhône, VI S 1/1.

47. "Cahier des charges pour l'entreprise et l'exécution d'un service régulière de correspondance par paquebots à vapeur entre Marseille, Cette et Toulon, et Alger, Oran, et Stora. Pendant douze années, du 1 Janvier 1854 à 31 Décembre 1865," Archives départementales des Bouches-du-Rhône, VI S 1/1.

48. See discussion of the explosion of a steamship boiler on the ship *La Ville de Nice*. Letter from the head engineer of the department, Marseille, February 6, 1864, Archives départementales des Bouches-du-Rhône, VI S 1/1.

49. Letter from the head engineer of the department, Marseille, February 6, 1864, 40.

50. Sewell, *Structure and Mobility*, 28.

51. On the development of steam-powered naval vessels, see, for example, Le Prince de Joinville, *Essais sur la Marine français* (Brussels: Meline, Cans, 1852).

52. See Édouard Paris's introduction to Pierre-Marie-Joseph de Bonnefoux and Édouard Paris, *Dictionnaire de marine à voiles et à vapeur*, 2nd ed. (Paris: Bouchard-Huzard, 1859), xiv.

53. De Bonnefoux and Paris, *Dictionnaire de marine à voiles et à vapeur*.

54. Statistic cited in Sewell, *Structure and Mobility*, 22.

55. James C. Williams, "Sailing as Play," *Icon* 19 (2013): 138.

56. Williams, "Sailing as Play," 173.

57. Jules Vence, *Construction et manoeuvre des bateaux et embarcations à voilure latine* (Paris: Augustin Challamel, 1897), 70.

58. Kate Brown and Thomas Klubock, "Environment and Labor: Introduction," in "Environment and Labor," ed. Brown and Klubock, special issue, *International Labor and Working-Class History*, no. 85 (Spring 2014): 6.

59. "Coup de vent du 11 février sur la Méditerranée," *Le Monde illustré* (Paris), February 25, 1865, 117.

Chapter Three

1. Henri Bouvier, [Directeur de l'École Normale d'Avignon] Giraud, and Alfred Pamard, *Le Mont-Ventoux* (Avignon: Seguin Frères,1879), 7.

2. Bouvier, Giraud, and Pamard, 12. The "zero degrees" refers to Celsius.

3. Bouvier, Giraud, and Pamard, 12.

4. Bouvier, Giraud, and Pamard, 16–17.

5. Alfred Pamard, *L'Observatoire du Mont-Ventoux: Communication faite à l'Académie de Vaucluse, le 7 Février 1918* (Avignon: François Séguin, 1918), 4.

6. Pamard, *L'Observatoire du Mont-Ventoux*, 12.

7. In addition to Mont Ventoux, the other "stations élevées" were located in Servance, Briançon, Puy de Dôme, Mont Aigoual, Mont Mounier, and the Pic du Midi. See Archives nationales, F17, box 3822.

8. Alfred de Vaulabelle, "Un nouvel observatoire: L'Observatoire du Mont Ventoux," *Magasin pittoresque*, no. 10 (1886): 164–66.

9. Eugène Barrême, "L'Observatoire météorologique du Ventoux," *Annales de Provence* 1 (1883): 149.

10. Charles Galtier, *Météorologie populaire dans la France ancienne: La Provence Empire du soleil et Royaume des vents* (Éditions Horvath, 1984), 142.

11. Petrarch quoted in Paul Pansier, "Les Ascensions du Ventoux et la Chapelle de la Ste-Croix du XIVe au XIXe siècle," *Annales d'Avignon et du Comtat Venaissin* 18 (1932): 138.

12. Martins, *Le Mont-Ventoux en Provence*, 15.

13. Pansier, "Les Ascensions du Ventoux," 145.

14. Complaint cited in Bernard Mondon and Steffen Lipp, *Petite anthologie du mistral* (Saint-Rémy-de-Provence: Éditions Équinoxe, 2004), 28.

15. Jean-Philippe Chassany, *Dictionnaire de météorologie populaire* (Paris: Maisonneuve et Larose, 1989), 231.

16. Letter from Urbain Jean-Joseph Le Verrier, Imperial Observatory of Paris, to the minister of public instruction, February 16, 1865, Archives nationales, F17, box 3727.

17. Letter from the Observatory of Paris to the prefect of the Vaucluse, March 5, 1879, Archives départementales de Vaucluse, 7 M 36.

18. Fabien Locher, *Le Savant et la tempête: Étudier l'atmosphère et prévoir le temps au XIXe siècle* (Rennes: Presses universitaires de Rennes, 2008), 37.

19. For an example of a standardized form sent from Paris to the French provinces, see the letter from the minister of public instruction, "Instructions pour les observations météorologiques des Écoles normales," August 8, 1865, Archives départementales de Vaucluse, 7 M 36.

20. See Vladimir Jankovic, *Reading the Skies: A Cultural History of English Weather, 1650–1820* (Chicago: University of Chicago Press, 2000); Jan Golinski, *British Weather and the Climate of Enlightenment* (Chicago: University of Chicago Press, 2007).

21. Lorraine Daston and Peter Galison, *Objectivity* (New York: Zone Books, 2007), 17.

22. Daston and Galison, *Objectivity*.

23. Locher, *Le Savant et la tempête*, 63.

24. Locher, 63.

25. Letter discussing the state of meteorological science in France, June 17, 1871, Archives départementales de Vaucluse, 7 M 36.

26. Circular from Jules Ferry, minister of public instruction, March 5, 1879, Paris, Archives départementales de Vaucluse, 7 M 36.

27. Barrême, "L'Observatoire météorologique du Ventoux," 187.

28. Félix Achard, *Une ascension au Ventoux* (Avignon: Imprimerie C. Maillet, 1875), 44.

29. Achard, *Une ascension au Ventoux*, 40.

30. See Michael S. Reidy, "Mountaineering, Masculinity, and the Male Body in Mid-Victorian Britain," *Osiris* 30, no. 1 (2015): 158–81.

31. On the links between masculinity and French citizenship in Third Republic France, see Judith Surkis, *Sexing the Citizen: Morality and Masculinity in France, 1870–1920* (Ithaca, NY: Cornell University Press, 2006).

32. Achard, *Une ascension au Ventoux*, 7.

33. Achard, 10.

34. Élie Margollé, "Commission météorologique du département de Vaucluse," *La Nature*, no. 240, January 5, 1878.

35. For a discussion of farmers' almanacs in nineteenth-century France, see Locher, *Le savant et la tempête*, 30.

36. Barrême, "L'Observatoire météorologique du Ventoux," 181.

37. Barrême.

38. Bouvier, Giraud, and Pamard, *Le Mont-Ventoux*, 13.

39. Deborah R. Coen, *Climate in Motion: Science, Empire, and the Problem of Scale* (Chicago: University of Chicago Press, 2018), 7–8.

40. Reclus, *Ocean, Atmosphere, and Life*, 207.

41. Bouvier, Giraud, and Pamard, *Le Mont-Ventoux*, 41.

42. See Coen, *Climate in Motion*, 11–12.

43. Barrême, "L'Observatoire météorologique du Ventoux," 187.

44. Bouvier, Giraud, and Pamard, *Le Mont-Ventoux*, 41.

45. Ministère de l'Agriculture, Direction des Forêts, Commission météorologique de Vaucluse, "Plan général de l'Observatoire," signed by M. Vincenti, notary, on January 18, 1886, Archives départementales de Vaucluse, 7 M 362.

46. Barrême, "L'Observatoire météorologique du Ventoux," 186.

47. See discussion on "the delicate and belated measurement of the speed of wind" in Martine Tabeaud et al., "Le risque 'coup de vent' en France depuis le XVIe siècle," *Annales de Géographie* 118, no. 667 (2009): 319.

48. Bill Streever, *And Soon I Heard a Rushing Wind: A Natural History of Moving Air* (New York: Little, Brown, 2016), 95.

49. Meteorological Commission of the Vaucluse, "Compte-rendu des Observations faites pendant l'année 1874," Archives départementales de Vaucluse, 7 M 38.

50. Meteorological Commission of the Vaucluse, "Compte-rendu des Observations faites pendant l'année 1874."

51. Gaston Tissander, "L'Observatoire météorologique du Mont Ventoux," *La Nature*, no. 599 (1884): 385.

52. De Vaulabelle, "Un nouvel observatoire," 164.

53. Inspection report on the Observatory of Mont Ventoux from the Central Meteorological Bureau, Paris, November 25, 1898, Archives nationales, F17, box 3727.

54. Tabeaud et al., "Le Risque 'coup de vent,' " 319.

55. Pamard, *L'Observatoire du Mont-Ventoux*, 16.

56. See Bruno Latour, *Science in Action: How to Follow Scientists and Engineers through Society* (Cambridge, MA: Harvard University Press, 1988).

57. "Rapport de la Commission nommée par le Ministre de l'Instruction publique pour étudier les voies et moyennes de développer la météorologie en France," ca. 1878, Archives nationales, F17, box 13592.

58. *Bulletin annuel de la commission du département des Bouches-du-Rhône, Année 1892* (Marseille: Typographie et lithographie Barlatier et Barthelet, 1893), 91.

59. "8 Décembre 1874. No. 1," Ministère de l'Instruction publique, Observatoire de Paris, VIII, *Avertissements météorologiques*, 1874, Archives nationales, F17, box 3727.

60. Locher, *Le Savant et la tempête*, 144.

61. See Mark Monmonier, *Air Apparent: How Meteorologists Learned to Map, Predict, and Dramatize the Weather* (Chicago: University of Chicago Press, 2000).

62. Robert Kohler, *Landscapes and Labscapes: Exploring the Lab-Field Border in Biology* (Chicago: University of Chicago Press, 2002), 8.

63. See agreement between the community of Bédoin and the prefect of the Vaucluse, August 29, 1879, Archives départementales de Vaucluse, 7 M 362.

64. Barrême, “L’Observatoire météorologique du Ventoux,” 187.

65. Letter from Henri Bouvier to the prefect, July 24, 1890, Archives départementales de Vaucluse, 7 M 36.

66. Letter from the Ingénieur ordinaire Lambert to the Service météorologique, May 21, 1895, Archives départementales de Vaucluse, 7 M 36.

67. Locher, *Le savant et la tempête*, 139.

68. Letter from the Ingénieur ordinaire to the Service météorologique, October 11, 1896, Archives départementales de Vaucluse, 7 M 36.

69. Inspection report on the Observatory of Mont Ventoux from the Central Meteorological Bureau, Paris, November 25, 1898, Archives nationales, F17, box 3727.

70. Letter from Henri Bouvier to the prefect, July 24, 1890, Archives départementales de Vaucluse, 7 M 36.

71. Letter from P. Provane to M. le Président de la Commission météorologique de Vaucluse, n.d., Archives départementales de Vaucluse, 7 M 36.

72. Letter from the head engineer to the prefect of the Vaucluse, December 4, 1908, Archives départementales de Vaucluse, 7 M 36.

73. Letter from Pierre Provane to the minister of public instruction, Bédoin, November 20, 1908, Archives départementales de Vaucluse, 7 M 36.

74. Letter from the head engineer to the prefect, October 25, 1915, Archives départementales de Vaucluse, 7 M 36.

75. Letter from the head engineer to the prefect, February 12, 1919, Archives départementales de Vaucluse, 7 M 36.

76. Letter from June 17, 1871, Archives départementales de Vaucluse, 7 M 36.

77. Letter from Henri Bouvier to the prefect of the Vaucluse, April 11, 1891, Archives départementales de Vaucluse, 7 M 36.

78. Pamard, *L’Observatoire du Mont-Ventoux*, 20.

Chapter Four

1. Dumas, *Pictures of Travel in the South of France*, 165.

2. Stendhal, *Mémoires d’un touriste*, vol. 1 (Paris: Michel Lévy Frères, 1854), 215.

3. “Avignon venteuse, avec vent ennuyeuse, sans vent pernicieuse.” Proverb cited in *Oeuvres de M. Michelet*, vol. 3 (Brussels: Meline, Cans, et Compagnie, 1840), 180.

4. “Quand le mistral entre par la fenêtre, le médecin sort par la porte.” Proverb cited in Jean-Paul Ferrier, *Leçons du territoire: Nouvelle géographie de la région Provence-Alpes-Côte d’Azur*, vol. 3 (Aix-en-Provence: Edisud, 1983), 38.

5. Historians of medicine and the environment have highlighted the significance of local topographies, ecologies, and climates in the perception of disease and public health in the nineteenth century. See, for example, Michael Osborne, *The Emergence of Tropical Medicine in France* (Chicago: University of Chicago Press, 2014); Elisabeth Hsu and Chris Low, eds., *Wind, Life, Health: Anthropological and Historical Perspectives* (Oxford: Blackwell, 2008); Linda Nash, *Inescapable Ecologies: A History of Environment, Disease, and Knowledge* (Berkeley: University of California Press, 2006); Conevery Bolton Valencius, *The Health of the Country: How American Settlers Understood Themselves and Their Land* (New York: Basic Books, 2002).

6. P.-A. Didiot, *Étude nouvelle du cholera: Historique, dynamique, prophylactique* (Marseille: L. Camoin, 1866), 27.

7. Letter from the Ministère de l'Instruction Publique, Édouard Lockroy, to the prefect of the Vaucluse, January 19, 1889, Archives départementales de Vaucluse, 7 M 36.

8. Geoffrey Lloyd, "Pneuma between Body and Soul," in *Wind, Life, Health*, ed. Hsu and Low, 132.

9. Lloyd, "Pneuma between Body and Soul," 132.

10. See Caroline C. Hannaway, "The Société Royale de Médecine and Epidemics in the Ancien Régime," *Bulletin of the History of Medicine* 46, no. 3 (May–June 1972): 257–73.

11. François Raymond, "Mémoire sur la topographie médicale de Marseille et son territoire; et sur celle des lieux voisins de cette ville," in Société royale de médecine, *Histoire de la Société royale de médecine* (Paris: Imprimerie de Philippe-Denys Pierres, 1780), 90.

12. Raymond, "Mémoire sur la topographie médicale de Marseille," 88.

13. [Louis] Bret d'Arles, "État des maladies qui ont reprises à Arles dans les diverses saisons de l'année 1783," Archives départementales des Bouches-du-Rhône, 1 J 980.

14. On the language that doctors used to describe malaria in the eighteenth century, see Emeline Roucaute et al., "Analysis of Causes of Spawning of Large-Scale, Severe Malarial Epidemics and Their Rapid Total Extinction in Western Provence, Historically a High Endemic Region of France (1745–1850), *Malaria Journal* 13, no. 72 (2014): 25.

15. Bret d'Arles, "État des maladies."

16. Michel Darluc, *Histoire naturelle de la Provence: Contenant ce qu'il y a de plus remarquable dans les règnes Végétal, Minéral, Animal & la partie Géoponique* (Avignon: J.J. Niel, 1782), 15–16.

17. Darluc, *Histoire naturelle de la Provence*, 260.

18. Salvatore Tresca (engraver) and Louis Lafitte (artist), *Ventôse* (Paris: Chez l'auteur, 1797).

19. Joseph Guérin, "Observations météorologiques de l'an XIII," Archives départementales de Vaucluse, 7 M 36.

20. Guérin, "Observations météorologiques de l'an XIII."

21. Joseph Guérin, "Météorologie, 1807: Tableau No. 1," Archives départementales de Vaucluse, 7 M 36.

22. Ulysse Denzel, *Coup d'oeil sur le dessèchement des marais de la Camargue* (Nîmes: Roger et Laporte, 1862), 4.

23. Bernard Picon, *L'Espace et le temps en Camargue: Histoire d'un delta face aux enjeux climatiques* (Arles: Actes Sud, 2020), 33.

24. Picon, *L'Espace et le temps en Camargue*.

25. For a longue durée study on the historical transformation of the Camargue, see Raphaël Mathevet et al., "Using Historical Political Ecology to Understand the Present: Water, Reeds, and Biodiversity in the Camargue Biosphere Reserve, Southern France," *Ecology and Society* 20, no. 4 (December 2015): 1–14, http://dx.doi.org/10.5751/ES-07787-200417.

26. François Poulle, *Étude de la Camargue ou statistique du delta du Rhône envisagé principalement sous le rapport des améliorations dont il est susceptible*, 1817, Museon Arlaten, CERCO, Pat-B-4890. Over the course of the following decades, Poulle would complete several more studies on the Camargue.

27. Poulle, *Étude de la Camargue*.

28. *Oxford English Dictionary Online*, s.v. "sirocco (*n.*)," revised 2002, https://www.oed.com/dictionary/sirocco_n.

29. Raymond Jaussaud, *Les Vents de Provence* (Berre: Association Sciences et Culture, 1991), 148.

30. John E. Oliver, ed., *Encyclopedia of World Climatology* (New York: Springer, 2008), 473.

31. Michael A. Osborne, *The Emergence of Tropical Medicine in France* (Chicago: University of Chicago Press, 2014). See also Michael A. Osborne, "The Geographic Imperative in Nineteenth-Century French Medicine," in *Medical Geography in Historical Perspective*, ed. Nicolaas A. Rupke, 31–50 (London: The Wellcome Trust for the History of Medicine at UCL, 2000).

32. Vladimir Jankovic, "Gruff Boreas, Deadly Calms: A Medical Perspective on Winds and the Victorians," *Journal of the Royal Anthropological Institute* 13, no. 1 (2007): 150.

33. Richard J. Evans, *Death in Hamburg: Society and Politics in the Cholera Years* (New York: Penguin Books, 2005), 22.

34. Statistic cited in Marc Aubert, "La Médecine à Marseille au XIXe siècle," *Provence historique* 43, no. 172 (1993): 195. By 1832, Marseille was Europe's third port in total maritime traffic, behind only London and Liverpool. Statistics from Osborne, *Emergence of Tropical Medicine in France*, 156.

35. Louis-E. Méry, *Le Choléra à Marseille: Seconde invasion, 1835* (Marseille: Feissat et Demonchy, 1836), 6.

36. Méry, *Le Choléra à Marseille*, 7.

37. Méry, 29.

38. Méry, 9.

39. For an overview of cholera's social and cultural dimensions in nineteenth-century France, see Catherine J. Kudlick, *Cholera in Post-Revolutionary Paris: A Cultural History* (Berkeley: University of California Press, 1996); David S. Barnes, *The Great Stink of Paris and the Nineteenth-Century Struggle against Filth and Germs* (Baltimore: Johns Hopkins University Press, 2006); Patrice Bourdelais and André Daudin, *Visages du choléra* (Paris: Belin, 1987).

40. Méry, *Le Choléra à Marseille*, 5.

41. Méry, 5.

42. Méry, 31.

43. Osborne, *Emergence of Tropical Medicine in France*, 47.

44. "Rapport sur une épidémie," Archives départementales des Bouches-du-Rhône, 5 M 159.

45. Aubert, "La Médecine à Marseille," 195.

46. Pierre-Auguste Didiot, *Étude nouvelle du choléra: Historique, dynamique, prophylactique* (Marseille: L. Camoin, 1866), 33.

47. Didiot, *Étude nouvelle du choléra*, 30.

48. Didiot, 25.

49. Didiot, 69.

50. Didiot, 27.

51. Reclus, *Ocean, Atmosphere, and Life*, 243.

52. Reclus, 243.

53. Bruno Martin, *Marseille: Précis monographique et encyclopédique ou le passé, le présent et l'avenir de cette ville* (Marseille: Imprimerie Samat, 1866), 13.

54. Osborne, *Emergence of Tropical Medicine in France*, 157.

55. Pierre-Auguste Didiot and Charles Guès, *Choléra épidémique de 1865: Rapports sur l'origine du choléra à Marseille en 1865* (Marseille: Imprimerie Samat, 1866), 58.

56. Martin, *Marseille*, 107.

57. Martin, 11.

58. Martine Tabeaud, "Climats urbains: Savoirs experts et pratiques sociales," *Ethnologie française* 40, no. 4 (October-December 2010): 688.

59. A. Daruty, *Le Mistral: Ses causes, ses effets, sa suppression* (Avignon: Imprimerie administrative de Seguin frères), 9.

60. Daruty, *Le Mistral*, 9.

61. Martin, *Marseille*, 11.

62. For the most up-to-date research on cholera transmission, see "Cholera–*Vibrio cholerae* Infection," Centers for Disease Control and Prevention, accessed May 22, 2023, https://www.cdc.gov/cholera/index.html.

63. For a comprehensive study on nineteenth-century industrial air pollution, see François Jarrige and Thomas Le Roux, *La contamination du monde: Une histoire des pollutions à l'âge industriel* (Paris: Seuil, 2017).

64. Marius Roux, *Rapport général des travaux des conseils d'hygiène et de salubrité des trois arrondissements: Département des Bouches-du-Rhône* (Marseille: Typographie Barlatier-Feissat et Demonchy, 1851), 17.

65. Martin, *Marseille*, 71.

66. Martin, 20–21.

67. See Denis Cosgrove, "Mappa Mundi, Anima Mundi: Imaginative Mapping and Environmental Representation," in *Ruskin and Environment: The Storm-Cloud of the Nineteenth Century*, ed. Michael Wheeler (Manchester: Manchester University Press, 1995), 9.

68. Vicky Albritton and Fredrik Albritton Jonsson, *Green Victorians: The Simple Life in John Ruskin's Lake District* (Chicago: University of Chicago Press, 2016), 36.

69. See Christine Corton, *London Fog: The Biography* (Cambridge, MA: Harvard University Press, 2015).

70. Xavier Daumalin, *Les Calanques industrielles de Marseille et leurs pollutions* (Aix-en-Provence: Ref.2c éditions, 2016), 67.

71. Daumalin, *Les Calanques industrielles de Marseille*, 67.

72. Daumalin.

73. Daumalin.

74. A. Clement, *Assainissement radical du port de Marseille* (Marseille: Typographie et Lithographie Ve Marius Olive, 1852), 4.

75. Clement, *Assainissement radical du port de Marseille*, 25.

76. Rodolphe Serre, *Port de Marseille: Ses eaux renouvelées en 36 heures* (Marseille: Petit Provençal, 1884), 8.

77. Serre, *Port de Marseille*, 9.

78. Serre, 9.

Chapter Five

1. On the natural artifacts embedded in plein air paintings, see Sopan Deb, "A Grasshopper Has Been Stuck in This van Gogh Painting for 128 Years," *New York Times*, November 8, 2017, https://www.nytimes.com/2017/11/08/arts/design/grasshopper-vincent-van-gogh-painting.html; Deborah Solomon, "Van Gogh and the Consolation of Trees," *New York Times*, May 14, 2023, https://www.nytimes.com/2023/05/11/arts/design/van-gogh-cypresses-met-museum.html.

2. For a discussion of the equipment used by nineteenth-century French landscape painters, see Anthea Callen, *The Work of Art: Plein Air Painting and Artistic Identity in Nineteenth-Century France* (London: Reaktion Books, 2015), 52–58.

3. On the concepts of "little" and "big" homelands in nineteenth-century France, see Anne-Marie Thiesse, *Ils apprenaient la France: L'Exaltation des régions dans le discours patriotique* (Paris: Éditions de la maison des sciences de l'homme, 1997); Stéphane Gerson, *The Pride of Place: Local Memories and Political Culture in Nineteenth-Century France* (Ithaca, NY: Cornell University Press, 2003).

4. Alessandro Nova, *The Book of the Wind: The Representation of the Invisible* (Montreal: McGill-Queen's University Press, 2011), 124–26.

5. Annette Haudiquet, Jacqueline Salmon, and Jean-Christian Fleury, "Le vent: 'Cela qui ne peut être peint,'" in *Le Vent: "Cela qui ne peut être peint,"* ed. Haudiquet, Salmon, and Fleury (Le Havre: MuMa, 2022), 13.

6. On the Barbizon School and the movement to preserve the Fontainebleau forest, see Ford, *Natural Interests*, 105–6.

7. On Loubon's evolution as a painter, see Jean-Roger Soubiran, *Le Paysage provençal et l'école de Marseille avant l'impressionnisme, 1845–1874* (Toulon: Réunion des Musées Nationaux, 1992), 197–215.

8. For a discussion of the natural subjects and groundbreaking painting techniques of the Barbizon circle of painters, see Scott Alan and Édouard Kopp, *Unruly Nature: The Landscapes of Théodore Rousseau* (Los Angeles: Getty Publications, 2016); Steven Adams, *The Barbizon School and the Origins of Impressionism* (New York: Phaidon Press, 1994).

9. Soubiran, *Le Paysage provençal*, 55.

10. Marie-Paule Vial, "L'École de Marseille," in *Sous le soleil, exactement: Le Paysage en Provence du classicisme à la modernité (1750–1920)*, ed. Guy Cogeval and Marie-Paule Vial (Montreal: Musée des Beaux-Arts de Montréal, 2005), 63. I would like to thank Marie-Paule Vial for meeting with me to discuss the role of the mistral in École de Marseille paintings.

11. Paule Brahic and Pierre Echinard, *La Provence rurale de 1850 à 1900 vue par ses peintres, ses félibres, et ses poètes* (Marseille: Fondation regards de Provence, 2005), 2.

12. Craig F. Bohren and Alistair B. Fraser, "Colors of the Sky," *Physics Teacher* 23, no. 5 (May 1985): 267–72.

13. Greg M. Thomas, "From Ecological Vision to Environmental Immersion: Theodore Rousseau to Claude Monet," in *From Corot to Monet: The Ecology of Impressionism*, ed. Stephen F. Eisenman (Rome: Skira, 2010), 47; see also Greg M. Thomas, *Art and Ecology in Nineteenth-Century France: The Landscapes of Théodore Rousseau* (Princeton, NJ: Princeton University Press, 2000).

14. Thomas, "From Ecological Vision to Environmental Immersion," 48.

15. Ford, *Natural Interests*, 95.

16. Callen, *The Work of Art*, 13.

17. Soubiran, *Le Paysage provençal*, 51.

18. Cogeval and Vial, *Sous le soleil*, xvii.

19. Jules Laforgue's 1883 description of impressionism, quoted in T. J. Clark, *The Painting of Modern Life: Paris in the Art of Manet and His Followers* (New York: Alfred A. Knopf, 1985), 8.

20. Richard R. Brettell, "Landscape, Nature, and 'La Nature,'" in *From Corot to Monet: The Ecology of Impressionism*, ed. Stephen F. Eisenman (Rome: Skira, 2010), 62.

21. For an account of Claude Monet's stay in Antibes, see Marianne Alphant, *Claude Monet: Une vie dans le paysage* (Paris: Hazan, 1993), 423–40.

22. Claude Monet, quoted in Alphant, *Claude Monet*, 438.

23. Stephen F. Eisenman explains that Impressionism's new approach to painting was the product of "a quick eye, a dexterous hand, and a sensitive temperament." See "From Corot to Monet: The Ecology of Impressionism," in Eisenman, *From Corot to Monet*, 17.

24. On the sky's role in landscape painting, see Tim Ingold, "The Eye of the Storm: Visual Perception and the Weather," *Visual Studies* 20, no. 2 (October 2005): 103–4.

25. Claude Monet, quoted in Alphant, *Claude Monet*, 436. For a discussion of the Impressionists' use of color, see Laura Anne Kalba, *Color in the Age of Impressionism: Commerce, Technology, and Art* (University Park, PA: Penn State University Press, 2017).

26. Callen, *The Work of Art*, 23.

27. Thomas, "From Ecological Vision to Environmental Immersion," 56.

28. Jules-Antoine Castagnary, quoted in Mary Tompkins Lewis, introduction to *Critical Readings in Impressionism and Post-Impressionism: An Anthology*, ed. Lewis (Berkeley: University of California Press, 2007), 2.

29. In recent years, the Impressionists' fascination with nature's sensory impacts on the human body has received attention from scientific researchers. See Aline Vedder et al., "Neurofunctional Correlates of Environmental Cognition: An fMRI Study with Images from Episodic Memory," *PLoS One* 10, no. 4 (April 2015): 1–11; Ernst Pöppel et al., "Sensory Processing of Art as a Unique Window into Cognitive Mechanisms: Evidence from Behavioral Experiments and fMRI Studies," *Procedia—Social and Behavioral Sciences* 86 (2013): 14.

30. See Douglas W. Druick and Peter Kort Zegers, *Van Gogh and Gauguin: The Studio of the South* (New York: Thames and Hudson, 2001).

31. On Gauguin's role as an ethnographer of "traditional" societies, see Stephen F. Eisenman, *Paul Gauguin: Artist of Myth and Dream* (Rome: Skira, 2008); Guillermo Solana, ed., *Gauguin and the Origins of Symbolism* (Madrid: Philip Wilson Publishers, 2004); Belinda Thomson, ed., *Gauguin: Maker of Myth* (London: Tate Publishing, 2010).

32. On Gauguin's and van Gogh's interests in the revival of traditional Provençal language, literature, ritual, and customs, see Debora Silverman, *Van Gogh and Gauguin: The Search for Sacred Art* (New York: Farrar, Straus and Giroux, 2000), 52–56.

33. Vincent van Gogh to Theo van Gogh, *Vincent van Gogh: The Letters*, letter 583, accessed June 27, 2019, http://vangoghletters.org/vg/letters/let583/letter.html.

34. Georges Roque, "Sous le soleil du Midi, la lumière devint couleur," in Françoise Cachin and Monique Nonne, *Méditerranée de Courbet à Matisse* (Paris: Réunion des musées nationaux, 2000), 115.

35. Vincent van Gogh to Émile Bernard, *Vincent van Gogh: The Letters*, letter 628, accessed June 27, 2019, http://vangoghletters.org/vg/letters/let628/letter.html.

36. Vincent van Gogh to Theo van Gogh, *Vincent van Gogh: The Letters*, letter 656, accessed June 27, 2019, http://vangoghletters.org/vg/letters/let656/letter.html.

37. Vincent van Gogh to Gabriel-Albert Aurier, *Vincent van Gogh: The Letters*, letter 853, accessed June 27, 2019, http://vangoghletters.org/vg/letters/let853/letter.html.

38. Tim Ingold uses van Gogh's paintings to argue that landscapes depend on movement to "come alive." See *Being Alive: Essays on Movement, Knowledge and Description* (London: Routledge, 2011), 122.

39. See Vojtech Jirat-Wasiutynski, "Vincent van Gogh's Paintings of Olive Trees and Cypresses from St.-Rémy," *Art Bulletin* 75, no. 4 (December 1993): 647–70.

40. Susan Alyson Stein, introduction to Stein, *Van Gogh's Cypresses* (New York: Metropolitan Museum of Art, 2023), 7.

41. Stein, *Van Gogh's Cypresses*, 7.

42. Results of a keyword search of *Vincent van Gogh: The Letters*, accessed July 10, 2023, https://vangoghletters.org/vg/search/simple?term=mistral.

43. Vincent van Gogh to Theo van Gogh, *Vincent van Gogh: The Letters*, letter 683, accessed June 27, 2019, http://vangoghletters.org/vg/letters/let683/letter.html.

44. Vincent van Gogh to Theo van Gogh, *Vincent van Gogh: The Letters*, letter 663, accessed June 27, 2019, https://vangoghletters.org/vg/letters/let663/letter.html.

45. Stendhal quoted in Laurence Wylie, *Village in the Vaucluse* (Cambridge, MA: Harvard University Press, 1974), 6.

46. See "Du Mistral (Son ancienneté): Explication du phénomène," in *Mémoires de l'Académie de Vaucluse* 9 (1890): 307.

47. See, for example, Jean-Pierre Besancenot, "Vents et santé en façade méditerranéenne de l'Europe," *Annales de Géographie* 98, no. 546 (March–April 1989): 179–95.

48. Nina Maria Athanassoglou-Kallmyer, *Cézanne and Provence: The Painter in His Culture* (Chicago: University of Chicago Press, 2003), 121.

49. Paul Cézanne, quoted in Athanassoglou-Kallmyer, *Cézanne and Provence*, 2.

50. Joachim Gasquet, *Cézanne* (Paris: Les Éditions Bernheim-Jeune, 1921), 60–61.

51. Gasquet, *Cézanne*, 20.

52. Athanassoglou-Kallmyer, *Cézanne and Provence*, 136.

53. Athanassoglou-Kallmyer, *Cézanne and Provence*, 171.

Epilogue

1. See LUMA Arles, "The Landscaped Park," accessed June 14, 2023, https://www.luma.org/en/arles/about-us/parc-des-ateliers/landscape-park.html.

2. Bas Smets and Eliane Leroux, "Luma Arles: The Making of a Climate," *TLmag36 Extended: All is Landscape*, January 21, 2022, https://tlmagazine.com/luma-arles-the-making-of-a-climate/.

3. Smets and Leroux, "Luma Arles."

4. For an overview of the project, see EDF, "Offshore Wind Power," accessed July 28, 2023, https://www.edf.fr/en/the-edf-group/producing-a-climate-friendly-energy/doubling-the-share-of-renewable-energies-by-2030/offshore-wind-power/we-are-preparing-the-offshore-wind-turbine-of-tomorrow.

5. See European Commission, "2030 Climate and Energy Framework," accessed July 22, 2021, https://ec.europa.eu/clima/policies/strategies/2030_en.

6. Anika Obermann et. al., "Mistral and Tramontane Wind Speed and Wind Direction Patterns for Regional Climate Simulations," *Climate Dynamics* 51, no. 3 (Aug. 2018): 1059–76; for a broader discussion on changing global wind speeds, see Zhenzhong Zeng et al., "A Reversal in Global Terrestrial Stilling and Its Implications for Wind Energy Production," *Nature Climate Change* 9 (December 2019): 979–85.

7. For a discussion of the multifaceted environmental threats facing contemporary Provence, see Julien Ruffault et al., "Daily Synoptic Conditions Associated with Large Fire Occurrence in Mediterranean France: Evidence for a Wind-Driven Fire Regime," *International Journal of Climatology* 37, no. 1 (2017): 524–33.

8. See Groupe régional d'experts sur le climat en Provence-Alpes-Côte d'Azur, *Climat et changement climatique en région Provence-Alpes-Côte d'Azur*, May 2016, 1–44.

9. David Glassberg, "Place, Memory, and Climate Change," *Public Historian* 36, no. 3 (2014): 17–30.

10. Georgina H. Endfield, "Exploring Particularity: Vulnerability, Resilience, and Memory in Climate Change Discourses," *Environmental History* 19, no. 2 (2014): 303–10.

BIBLIOGRAPHY

Abulafia, David. *The Great Sea: A Human History of the Mediterranean*. Oxford: Oxford University Press, 2011.

Achard, Félix. *Une ascension au Ventoux*. Avignon: Imprimerie C. Maillet, 1875.

Adams, Steven. *The Barbizon School and the Origins of Impressionism*. New York: Phaidon Press, 1994.

Alan, Scott, and Édouard Kopp. *Unruly Nature: The Landscapes of Théodore Rousseau*. Los Angeles: Getty Publications, 2016.

Albritton, Vicky, and Fredrik Albritton Jonsson. *Green Victorians: The Simple Life in John Ruskin's Lake District*. Chicago: University of Chicago Press, 2016.

Allen, Harriet D. *Mediterranean Ecogeography*. London: Pearson Press, 2001.

Alphant, Marianne. *Claude Monet: Une vie dans le paysage*. Paris: Hazan, 1993.

Amélineau, Françoise, Clara Péron, Amélie Lescroël, Matthieu Authier, Pascal Provost, and David Grémillet. "Windscape and Tortuosity Shape the Flight Costs of Northern Gannets." *Journal of Experimental Biology* 217, no. 6 (March 2014): 876–85.

Anderson, Benedict. *Imagined Communities: Reflections on the Origin and Spread of Nationalism*. London: Verso, 2016.

Anderson, Katharine. *Predicting the Weather: Victorians and the Science of Meteorology*. Chicago: University of Chicago Press, 2005.

Antier, Jean-Jacques. *Marins de Provence et du Languedoc: Vingt-cinq siècles d'histoire du littoral français méditerranéen*. Montpellier: Les Presses de Languedoc, 2003.

Athanassoglou-Kallmyer, Nina Maria. *Cézanne and Provence: The Painter in His Culture*. Chicago: University of Chicago Press, 2003.

Aubanel, Théodore. *Les Filles d'Avignon*. Paris: Nouvelle Librairie Parisienne, 1891.

Aubert, Marc. "La Médecine à Marseille au XIXe siècle." *Provence historique* 43, no. 172 (1993): 185–206.

Barnes, David S. *The Great Stink of Paris and the Nineteenth-Century Struggle against Filth and Germs*. Baltimore: Johns Hopkins University Press, 2006.

Barnett, Cynthia. *Rain: A Natural and Cultural History*. New York: Penguin Press, 2015.

Barrême, Eugène. "L'Observatoire météorologique du Ventoux." *Annales de Provence* 1 (1883).

Becquerel, Antoine-César. *Des climats et de l'influence qu'exercent les sols boisés et non boisés*. Paris: Didot Frères, 1853.

Behringer, Wolfgang. *A Cultural History of Climate*. Translated by Patrick Camiller. Cambridge: Polity Press, 2010.

Bennett, Jane. *Vibrant Matter: A Political Ecology of Things*. Durham, NC: Duke University Press, 2010.

Benoit, Fernand. *La Provence et le Comtat Venaissin: Arts et traditions populaires*. Avignon: Aubanel, 1975.

Berthelot, Sabin. *Étude sur les pêches maritimes dans la méditerranée et l'océan*. Paris: Challamel Ainé, 1868.

Besancenot, Jean-Pierre. "Vents et santé en façade méditerranéenne de l'Europe." *Annales de Géographie* 98, no. 546 (March-April 1989): 179–95.

Bess, Michael. *The Light-Green Society: Ecology and Technological Modernity in France, 1960–2000*. Chicago: University of Chicago Press, 2003.

Blondel, Jacques, and James Aronson. *Biology and Wildlife of the Mediterranean Region*. Oxford: Oxford University Press, 1999.

Bohren, Craig F., and Alistair B. Fraser. "Colors of the Sky." *Physics Teacher* 23, no. 5 (May 1985): 267–72.

Boia, Lucian. *The Weather in the Imagination*. London: Reaktion Books, 2005.

Bonneuil, Christophe, and Jean-Baptiste Fressoz. *L'Évènement anthropocène: La Terre, l'histoire, et nous*. Paris: Points, 2016.

Bourdelais, Patrice, and André Daudin. *Visages du Choléra*. Paris: Belin, 1987.

Bourrilly, J., and Louis Aubert, *Objets et rites talismaniques en Provence d'après les collections du Museon Arlaten*. Valence: Imprimerie Valentinoise, 1907.

Bouvier, Henri, [Directeur de l'École Normale d'Avignon] Giraud, and Alfred Pamard. *Le Mont-Ventoux*. Avignon: Seguin Frères, 1879.

Bouvier, Henri, and Alfred Pamard. *Notice sur l'observatoire du Mont-Ventoux*. Avignon: Seguin Frères, 1884.

Boyer, F., A. Orieux, and E. Pouget. *Le Mistral en Provence occidental*. Paris: Météorologie nationale monographie, 1970.

Brahic, Paule, and Pierre Echinard, eds. *La Provence rurale de 1850 à 1900 vue par ses peintres, ses félibres, et ses poètes*. Marseille: Fondation regards de Provence, 2005.

Braudel, Fernand. *The Identity of France*. Vol. 1. Translated by Siân Reynolds. New York: Harper and Row, 1988.

Braudel, Fernand. *The Mediterranean and the Mediterranean World in the Age of Philip II*. Vol. 1. Translated by Siân Reynolds. Berkeley: University of California Press, 1995.

Braudel, Fernand. *Memory and the Mediterranean*. Translated by Siân Reynolds. New York: Vintage, 2002.

Brettell, Richard R. "Landscape, Nature, and 'La Nature.'" In *From Corot to Monet: The Ecology of Impressionism*, edited by Stephen F. Eisenman, 59-69. Rome: Skira, 2010.

Brown, Kate, and Thomas Klubock. "Environment and Labor: Introduction." *International Labor and Working-Class History* 85 (Spring 2014): 4–9.

Bulletin annuel de la commission du département des Bouches-du-Rhône, année 1892. Marseille: Typographie et lithographie Barlatier et Barthelet, 1893.

Cachin, Françoise, and Monique Nonne. *Méditerranée de Courbet à Matisse*. Paris: Réunion des musées nationaux, 2000.

Callen, Anthea. *The Work of Art: Plein Air Painting and Artistic Identity in Nineteenth-Century France*. London, 2015.

Campaignac, A. *De l'état actuel de la navigation par la vapeur*. Paris: Librairie industrielle-scientifique, 1842.

Carter, Thomas, and Elizabeth Collins Cromley. *Invitation to Vernacular Architecture: A Guide to the Study of Ordinary Buildings and Landscapes*. Knoxville: University of Tennessee Press, 2005.

Castagne, Louis. *Observations sur le reboisement des montagnes et des terrains vagues, dans le département des Bouches-du-Rhône*. Aix-en-Provence: Vve Taverin, 1851.

Cazalat, Galy. *Mémoire théorique et pratique sur les bateaux à vapeur*. Paris: Librairie scientifique et industrielle de L. Mathias, 1837.

Caziot, Eugène. "Du Mistral (Son ancienneté): Explication du phénomène." *Mémoires de l'Académie de Vaucluse* 9 (1890): 303–30.

Centers for Disease Control and Prevention. "Cholera–*Vibrio cholerae* Infection." Last modified November 14, 2022. https://www.cdc.gov/cholera/index.html.

Chanet, Jean-François. *L'École républicaine et les petites patries*. Paris: Aubier, 1996.

Chassany, Jean-Philippe. *Dictionnaire de météorologie populaire*. Paris: Maisonneuve et Larose, 1989.

Chevalier, Michel. *Essais de politique industrielle*. Paris: Librairie de Charles Gosselin, 1843.

Cioc, Mark. *The Rhine: An Eco-Biography, 1815–2000*. Seattle: University of Washington Press, 2002.

Clancy-Smith, Julia. *Mediterraneans: North Africa and Europe in the Age of Migration, c. 1800–1900*. Berkeley: University of California Press, 2011.

Clark, T. J. *The Painting of Modern Life: Paris in the Art of Manet and His Followers*. New York: Alfred A. Knopf, 1985.

Claval, Paul. "From Michelet to Braudel: Personality, Identity and Organization of France." In *Geography and National Identity*, edited by David Hooson, 39–57. Oxford: Blackwell, 1994.

Clement, A. *Assainissement radical du port de Marseille*. Marseille: Typographie et Lithographie Ve Marius Olive, 1852.

Cobb, Rachel. *Mistral: The Legendary Wind of Provence*. Bologna: Damiani Press, 2018.

Coen, Deborah R. "Climate and Circulation in Imperial Austria." *Journal of Modern History* 82 (2010): 839–75.

Coen, Deborah R. *Climate in Motion: Science, Empire, and the Problem of Scale*. Chicago: University of Chicago Press, 2018.

Coen, Deborah R. "Imperial Climatologies from Tyrol to Turkestand." *Osiris* 26 (2011): 45–65.

Cogeval, Guy, and Marie-Paule Vial, eds. *Sous le soleil, exactement: Le Paysage en Provence du classicisme à la modernité*. Montreal: Montreal Museum of Fine Arts, 2005.

Conrad, Joseph. *The Arrow of Gold: A Story between Two Notes*. New York: Doubleday, 1921.

Corbin, Alain. *Le Ciel et la mer*. Paris: Bayard, 2005.

Corbin, Alain. *The Lure of the Sea: The Discovery of the Seaside in the Western World, 1750–1840*. Translated by Jocelyn Phelps. New York: Penguin Press, 1995.

Corbin, Alain. *La Rafale et le zéphyr: Histoire des manières d'éprouver et de rêver le vent*. Paris: Fayard, 2021.

Corton, Christine. *London Fog: The Biography*. Cambridge, MA: Belknap Press, 2015.

Corvol, Andrée. *Les Sources de l'histoire de l'environnement: Le XIXe siècle*. Paris: L'Harmattan, 1999.

Cosgrove, Denis. "Mappa Mundi, Anima Mundi: Imaginative Mapping and Environmental Representation." In *Ruskin and the Environment: The Storm-Cloud of the Nineteenth*

Century, edited by Michael Wheeler. Manchester: Manchester University Press, 1995.

"Coup de vent du 11 février sur la Méditerranée." *Le monde illustré* (Paris), February 25, 1865.

Cronon, William. "A Place for Stories: Nature, History, and Narrative." *Journal of American History* 78, no. 4 (March 1992): 1347–76.

Crosby, Alfred W. "Winds." In *Ecological Imperialism: The Biological Expansion of Europe, 900–1900*. Cambridge: Cambridge University Press, 2004.

Darluc, Michel. *Histoire naturelle de Provence: Contenant ce qu'il y a de plus remarquable dans les règnes Végétal, Minéral, Animal & la partie Géoponique*. Avignon: J. J. Niel, 1792.

Darnton, Robert. *The Great Cat Massacre and Other Episodes in French Cultural History*. New York: Vintage Books, 1985.

Daruty, A. *Le Mistral: Ses causes, ses effets et sa suppression*. Avignon: Imprimerie administrative de Seguin frères, 1879.

Dassié, Véronique, and Dominique Séréna-Allier. "Are Popular Local Artefacts Exotic? Building 'Provençalness' at the Museon Arletan [*sic*]." *Journal of Museum Ethnography*, no. 22 (December 2009): 129–41.

Daston, Lorraine. "Unruly Weather: Natural Law Confronts Natural Variability." In *Natural Law and the Laws of Nature in Early Modern Europe: Jurisprudence, Theology, Moral and Natural Philosophy*, edited by Lorraine Daston and Michael Stolleis, 233–48. Burlington, VT: Ashgate, 2008.

Daston, Lorraine, and Peter Galison. *Objectivity*. New York: Zone Books, 2007.

Daston, Lorraine, and Elizabeth Lunbeck, eds. *Histories of Scientific Observation*. Chicago: University of Chicago Press, 2011.

Daudet, Alphonse. *Letters from My Mill*. Translated by Katharine Prescott Wormeley. Boston: Little, Brown, 1900.

Daudet, Alphonse. *Letters from My Windmill*. Translated by Dédicace. Rennes: Éditions Ouest-France, 2011.

Daudet, Alphonse. *Tartarin of Tarascon*. Boston: T. Y. Crowell, 1895.

Daumalin, Xavier. *Les Calanques industrielles de Marseille et leurs pollutions*. Aix-en-Provence: Ref.2c éditions, 2016.

Davin, Emmanuel. *Mistral et autres vents en Provence*. Toulon: Société nouvelle des imprimeries Toulonnaises, 1938.

Davis, Diana. *The Arid Lands: History, Power, Knowledge*. Cambridge, MA: MIT Press, 2016.

Deb, Sopan. "A Grasshopper Has Been Stuck in This van Gogh Painting for 128 Years." *New York Times*, November 8, 2017. https://www.nytimes.com/2017/11/08/arts/design/grasshopper-vincent-van-gogh-painting.html.

DeBlieu, Jan. *Wind: How the Flow of Air Has Shaped Life, Myth, and the Land*. Boston: Houghton Mifflin, 1998.

de Bonnefoux, Pierre-Marie-Joseph, and Édouard Paris. *Dictionnaire de marine à voiles et à vapeur*. Paris: Arthus Bertrand, 1859.

de Certeau, Michel, Dominique Julia, and Jacques Revel. *Une politique de la langue: La Révolution française et les patois; L'Enquête de Grégoire*. Paris: Gallimard, 1975.

DeGroot, Dagomar. *The Frigid Golden Age: Climate Change, the Little Ice Age, and the Dutch Republic, 1560–1720*. Cambridge: Cambridge University Press, 2018.

Delamarre, L.-G. *Traité pratique de la culture des pins*. Paris: Chez Madame Huzard, 1831.

Demontzey, Prosper. *Traité pratique du reboisement et du gazonnement des montagnes*. Paris: J. Rothschild, 1882.

Denzel, Ulysse. *Coup d'oeil sur le dessèchement des marais de la Camargue*. Nîmes: Roger et Laporte, 1862.

De Saussure, Horace-Bénédict. *Voyages dans les Alpes*. Neuchâtel: Louis Fauche-Borel, 1796.

Didion, Joan. "Los Angeles Notebook." In *Slouching Towards Bethlehem*. New York: Farrar, Straus and Giroux, 1968.

Didiot, P.-A. *Étude nouvelle du choléra: Historique, dynamique, prophylactique*. Marseille: L. Camoin, 1866.

Druick, Douglas W., and Peter Kort Zegers. *Van Gogh and Gauguin: The Studio of the South*. New York: Thames and Hudson, 2001.

Duffy, Andrea E. *Nomad's Land: Pastoralism and French Environmental Policy in the Nineteenth-Century Mediterranean World*. Lincoln: University of Nebraska Press, 2019.

Dufour, L. "Le Mistral dans la littérature française." *Ciel et Terre* 94 (1978): 34–37.

Dumas, Alexandre. *Pictures of Travel in the South of France*. London: Offices of the National Illustrated Library, 1851.

Dunlop, Catherine T. *Cartophilia: Maps and the Search for Identity in the French-German Borderland*. Chicago: University of Chicago Press, 2015.

Dunlop, Catherine T. "Looking at the Wind: Paintings of the Mistral in Fin-de-Siècle France." *Environmental History* 20 (2015): 505–18.

Dunlop, Catherine T. "Losing an Archive: Doing Place-Based History in the Age of the Anthropocene." *American Historical Review* 126, no. 3 (September 2021): 1143–53.

EDF. "Offshore Wind Power." Accessed July 28, 2023. https://www.edf.fr/en/the-edf-group/producing-a-climate-friendly-energy/doubling-the-share-of-renewable-energies-by-2030/offshore-wind-power/we-are-preparing-the-offshore-wind-turbine-of-tomorrow.

Eisenman, Stephen F., ed. *From Corot to Monet: The Ecology of Impressionism*. Rome: Skira, 2010.

Eisenman, Stephen F., ed. *Paul Gauguin: Artist of Myth and Dream*. Rome: Skira, 2008.

Endfield, Georgina H. "Exploring Particularity: Vulnerability, Resilience, and Memory in Climate Change Discourses." *Environmental History* 19, no. 2 (April 2014): 303–10.

European Environmental Agency. "Biogeographical Regions in Europe." June 19, 2017. https://www.eea.europa.eu/data-and-maps/figures/biogeographical-regions-in-europe-2.

European Commission. "2030 Climate and Energy Framework." Accessed July 22, 2021. https://ec.europa.eu/clima/policies/strategies/2030_en.

Evans, Richard J. *Death in Hamburg: Society and Politics in the Cholera Years*. New York: Penguin Books, 2005.

Fagan, Brian. *The Little Ice Age: How Climate Made History, 1300–1850*. New York: Basic Books, 2000.

Faget, Daniel. *Marseille et la mer: Hommes et environnement marin (XVIIe-XXe siècle)*. Rennes: Presses universitaires de Rennes, 2011.

Ferrier, Jean-Paul. *Leçons du territoire, nouvelle géographie de la région Provence-Alpes-Côte d'Azur*. 3 vols. Aix-en-Provence: Edisud, 1983.

Fiege, Mark. *Irrigated Eden: The Making of an Agricultural Landscape in the American West*. Seattle: University of Washington Press, 1999.

Finch-Race, Daniel, and Valentina Gosetti. "Editorial: Discovering Industrial-Era French Ecoregions." *Dix-Neuf* 23, no. 3–4 (2019): 151–62.

Fleming, James Rodger. *Inventing Atmospheric Science: Bjerknes, Rossby, Wexler, and the Foundations of Modern Meteorology*. Cambridge, MA: MIT Press, 2016.

Fleming, James Rodger, Vladimir Jankovic, and Deborah R. Coen, eds. *Intimate Universality: Local and Global Themes in the History of Weather and Climate*. Sagamore Beach, MA: Science History Publications, 2006.

Ford, Caroline. *Natural Interests: The Struggle over Environment in Modern France*. Cambridge, MA: Harvard University Press, 2016.

Ford, Caroline. "Nature's Fortunes: New Directions in the Writing of European Environmental History." *Journal of Modern History* 79 (March 2007): 112–33.

Ford, Caroline, and Tamara L. Whited, eds. "New Directions in French Environmental History." Special issue, *French Historical Studies* 32, no. 3 (Summer 2009).

Fowle, Francis, and Richard Thompson, eds. *Soil and Stone: Impressionism, Urbanism, and Environment*. Burlington, VT: Ashgate, 2003.

Fressoz, Jean-Baptiste. *L'Apocalypse joyeuse: Une histoire du risque technologique*. Paris: Seuil, 2012.

Fressoz, Jean-Baptiste, and Fabien Locher. *Les Révoltes du ciel: Une histoire du changement climatique XVe–XXe siècle*. Paris: Seuil, 2020.

Gade, Daniel W. "Windbreaks in the Lower Rhône Valley." *Geographical Review* 68, no.7 (April 1978): 127–44.

Galtier, Charles. *Météorologie populaire dans la France ancienne: La Provence, Empire du soleil et Royaume des vents*. Éditions Horvath, 1984.

Gasquet, Joachim. *Cézanne*. Paris: Éditions Bernheim-Jeune, 1921.

Gayford, Martin. *The Yellow House: Van Gogh, Gauguin, and Nine Turbulent Weeks in Provence*. New York: First Mariner Books, 2008.

Gelu, Victor. *Oeuvres complètes de Victor Gelu*. 2 vols. Paris: G. Charpentier, 1886.

Gerson, Stéphane. *The Pride of Place: Local Memories and Political Culture in Nineteenth-Century France*. Ithaca, NY: Cornell University Press, 2003.

Glacken, Clarence. *Traces on the Rhodian Shore*. Berkeley: University of California Press, 1967.

Glassberg, David. "Place, Memory, and Climate Change." *Public Historian* 36, no. 3 (August 2014): 17–30.

Golinski, Jan. *British Weather and the Climate of Enlightenment*. Chicago: University of Chicago Press, 2007.

Gouirand, André. *Les Peintres provençaux: Loubon et son temps*. Paris: Société d'éditions littéraires et artistiques, 1901.

Granger, Paul. *Les Fleurs du Midi: Cultures florales industrielles*. Paris: Librairie J.-B. Baillière et fils, 1902.

Groupe régional d'experts sur le climat en Provence-Alpes-Côte d'Azur. *Climat et changement climatique en région Provence-Alpes-Côte d'Azur* (May 2016): 1–44.

Grove, A. T., and Oliver Rackham. *The Nature of Mediterranean Europe: An Ecological History*. New Haven, CT: Yale University Press, 2003.

Gueusquin, Marie-France. "Des vents, des espaces et des hommes (Provence, Cotentin, Flandre)." *Études rurales* 177 (January-June 2006): 121–36.

Hall, Marcus. *Earth Repair: A Transatlantic History of Environmental Restoration*. Charlottesville: University of Virginia Press, 2005.

Hamilton, Sarah R. *Cultivating Nature: The Conservation of a Valencian Working Landscape*. Seattle: University of Washington Press, 2018.

Hannaway, Caroline C. "The Société Royale de Médicine and Epidemics in the Ancien Régime." *Bulletin of the History of Medicine* 46, no. 3 (May-June 1972): 257–73.

Harley, J. B. *The New Nature of Maps: Essays in the History of Cartography*. Baltimore: Johns Hopkins University Press, 2001.

Harp, Stephen L. *The Riviera, Exposed: An Ecohistory of Postwar Tourism and North African Labor*. Ithaca, NY: Cornell University Press, 2022.

Harris, Alexandra. *Weatherland: Writers and Artists under English Skies*. London: Thames and Hudson, 2016.

Harvey, David. *Paris: Capital of Modernity*. New York: Routledge, 2005.

Haudiquet, Annette, Jacqueline Salmon, and Jean-Christian Fleury, eds. *Le Vent: "Cela qui ne peut être peint."* Le Havre: MuMa, 2022.

Hayes-Conroy, Jessica, and Allison Hayes-Conroy. "Visceral Geographies: Mattering, Relating, and Defying." *Geography Compass* 4, no. 9 (2010): 1273–83.

Homet, Jean-Marie. *Provence des moulins à vent*. Aix-en-Provence: Edisud, 1984.

Horowitz, Helen Lefkowitz. *A Taste for Provence*. Chicago: University of Chicago Press, 2016.

Hughes, Donald J. *The Mediterranean: An Environmental History*. Santa Barbara: ABC Clio, 2005.

Hunt, Nick. *Where the Wild Winds Are: Walking Europe's Winds from the Pennines to Provence*. London: Nicholas Brealey Publishing, 2017.

Hsu, Elisabeth, and Chris Low, eds. *Wind, Life, Health: Anthropological and Historical Perspectives*. Oxford: Blackwell, 2008.

Ingold, Tim. *Being Alive: Essays on Movement, Knowledge, and Description*. London: Routledge, 2011.

Ingold, Tim. "The Eye of the Storm: Visual Perception and the Weather." *Visual Studies* 20, no. 2 (October 2005): 97–104.

Ingold, Tim. "Footprints through the Weather World: Walking, Breathing, Knowing." *Journal of the Royal Anthropological Institute* 16, no. 1 (May 2010): 121–39.

Ingold, Tim. *The Life of Lines*. New York: Routledge, 2015.

Ingold, Tim. *The Perception of the Environment: Essays on Livelihood, Dwelling, and Skill*. New York: Routledge, 2011.

Jackson, Jeffrey H. *Paris under Water: How the City of Light Survived the Great Flood of 1910*. New York: St. Martin's Griffin, 2010.

Jacq, Valérie, Philippe Albert and Robert Delorme. "Le Mistral: Quelques aspects des connaissances actuelles." *La Météorologie* 50 (August 2005): 30–38.

Jankovic, Vladimir. "Gruff Boreas, Deadly Calms: A Medical Perspective on Wind and the Victorians." *Journal of the Royal Anthropological Institute* 13, no. 1 (April 2007): 147–64.

Jankovic, Vladimir. "The Place of Nature and the Nature of Place: The Chorographic Challenge to the History of British Provincial Science," *History of Science* 38 (2000): 79–113.

Jankovic, Vladimir. *Reading the Skies: A Cultural History of English Weather, 1650–1820*. Chicago: University of Chicago Press, 2000.

Jansen, Leo, Hans Luijten, and Nienke Bakker, eds. *Vincent van Gogh: The Letters*. Amsterdam and The Hague: Van Gogh Museum and Huygens ING, 2009. http://vangoghletters.org.

Jarrige, François, and Thomas Le Roux. *La Contamination du monde: Une histoire des pollutions à l'âge industriel*. Paris: Seuil, 2017.

Jaubert, J.-B., and Barthélemy-Lapommeraye. *Richesses ornithologiques du midi de la France: Description méthodique de tous les oiseaux observés en Provence et dans les départements circonvoisins*. Marseille: Barlatier-Feissat et Demonchy, 1859.

Jaussaud, Raymond. *Les Vents de Provence*. Berre: Association Sciences et Culture, 1991.

Jirat-Wasiutynski, Vojtech. "Vincent van Gogh's Paintings of Olive Trees and Cypresses from St.-Rémy." *Art Bulletin* 75, no. 4 (December 1993): 647–70.

Joinville, Le Prince de. *Essais sur la Marine français*. Brussels: Meline, Cans, 1852.

Jones, Peter M. *Agricultural Enlightenment: Knowledge, Technology, and Nature, 1750–1840*. Oxford: Oxford University Press, 2016.

Jouffroy d'Abbans, Achille-François-Léonor de. *Des bateaux à vapeur: Précis historique de leur invention*. Paris: Imprimerie de E. Duvenger, 1839.

Jouven, Alphonse. *Sur les moulins à farine*. Aix-en-Provence: Imprimerie de Nicot et Pardigon, 1848.

Julliany, Jules. *Essai sur le commerce de Marseille*. Marseille: Jules Barile, 1842.

Kalba, Laura Anne. *Color in the Age of Impressionism: Commerce, Technology, and Art*. University Park, PA: Penn State University Press, 2017.

Kaplan, Steven L. *Provisioning Paris: Merchants and Millers in the Grain and Flour Trade during the Eighteenth Century*. Ithaca, NY: Cornell University Press, 1984.

Keller, Richard C. *Fatal Isolation: The Devastating Paris Heat Wave of 2003*. Chicago: University of Chicago Press, 2015.

Kohler, Robert E. *Landscapes and Labscapes: Exploring the Lab-Field Border in Biology*. Chicago: University of Chicago Press, 2002.

Krishnan, Siddhartha, Christopher L. Pastore, and Samuel Temple. "Introduction: Unruly Matters." In "Unruly Environments," edited by Siddhartha Krishnan, Christopher L. Pastore, and Samuel Temple. Special issue, *Rachel Carson Center Perspectives* 3 (2015): 5–7.

Kudlick, Catherine. *Cholera in Post-Revolutionary Paris: A Cultural History*. Berkeley: University of California Press, 1996.

Latour, Bruno. *Science in Action: How to Follow Scientists and Engineers through Society*. Cambridge, MA: Harvard University Press, 1988.

Laure, Henri. *Manuel du cultivateur provençal ou Cours d'Agriculture simplifiée pour le midi de l'Europe et le Nord de l'Afrique*. Toulon: Chez Teisseire, 1839.

LeCain, Timothy J. *The Matter of History: How Things Create the Past*. Cambridge: Cambridge University Press, 2017.

Leméthéyer, François-Frédéric. *Dictionnaire moderne des termes de marine et de la navigation à vapeur*. Paris: Robiquet, 1843.

Le Roy Ladurie. *Times of Feast, Times of Famine: A History of Climate since the Year 1000*. New York: Doubleday, 1971.

Lewis, Jayne. *Air's Appearance: Literary Atmosphere in British Fiction*. Chicago: University of Chicago Press, 2012.

Lewis, Martin W., and Kären E. Wigen. *The Myth of Continents: A Critique of Metageography*. Berkeley: University of California Press, 1997.

Lewis, Mary Tompkins, ed. *Critical Readings in Impressionism and Post-Impressionism: An Anthology*. Berkeley: University of California Press, 2007.

Locher, Fabien. "Le Rentier et le baromètre: Météorologie 'savante' et météorologie 'profane' au XIXe siècle." *Ethnologie française* 39, no. 4 (2009): 645–53.

Locher, Fabien. *Le Savant et la tempête: Étudier l'atmosphère et prévoir le temps au XIXe siècle*. Rennes: Presses universitaires de Rennes, 2008.

Locher, Fabien, and Jean-Baptiste Fressoz. "Modernity's Frail Climate: A Climate History of Environmental Reflexivity." *Critical Inquiry* 38, no. 3 (Spring 2012): 579–98.

LUMA Arles. "The Landscaped Park." Accessed June 14, 2023. https://www.luma.org/en/arles/about-us/parc-des-ateliers/landscape-park.html.

Maher, Neil M. "Body Counts: Tracking the Human Body through Environmental History." In *A Companion to American Environmental History*, edited by Douglas Cazaux Sackman, 163–80. Oxford: Blackwell, 2014.

Margollé, Élie. "Commission météorologique du département de Vaucluse." *La Nature*, no. 240. January 5, 1878.

Marsh, George Perkins. *Man and Nature; or, Physical Geography as Modified by Human Action*. London: Sampson Low, Son and Marsten, 1864.

Martel, Philippe. *Félibres et leur temps: Renaissance d'oc et opinion, 1850–1914*. Pessac: Presses universitaires de Bordeaux, 2010.

Martin, Bruno. *Marseille: Précis monographique et encyclopédique ou le passé, le présent et l'avenir de cette ville*. Marseille: Imprimerie Samat, 1866.

Martins, Charles. *Le Mont-Ventoux en Provence*. Paris: Imprimerie de J. Claye, 1863.

Marx, Karl. *Capital*. Translated by Samuel Moore and Edward Avelig. Vol. 1. New York: International Publishers, 1992.

Massot, Jean-Luc. *Maisons rurales et vie paysanne en Provence*. Paris: Serg-Berger-Levrault, 1992.

Matagne, Patrick. *Aux origines de l'écologie: Les Naturalistes en France de 1800 à 1914*. Paris: Éditions du CTHS, 1999.

Mathevet, Raphaël, Nancy Lee Peluso, Alexandre Couespel, and Paul Robbins. "Using Historical Political Ecology to Understand the Present: Water, Reeds, and Biodiversity in the Camargue Biosphere Reserve, Southern France." *Ecology and Society* 20, no. 4 (December 2015): 1–14. http://dx.doi.org/10.5751/ES-07787-200417.

Matteson, Kieko. *Forests in Revolutionary France: Conservation, Community, and Conflict, 1669–1848*. Oxford: Oxford University Press, 2015.

Mauro, Giovanni. "The New 'Windscapes' in the Time of Energy Transition: A Comparison of Ten European Countries." *Applied Geography*, 109 (2019). https://doi.org/10.1016/j.apgeog.2019.102041.

McHarg, Ian. *Design with Nature*. New York: Natural History Press, 1969.

McNeill, J. R. *The Mountains of the Mediterranean World: An Environmental History*. Cambridge: Cambridge University Press, 2003.

McPhee, Peter. *Revolution and Environment in Southern France, 1780–1830*. Oxford: Oxford University Press, 1999.

Meissonier, Jean. *Voiliers de l'époque romantique peints par Antoine Roux*. Lausanne: Edita Lausanne, 1968.

Mercier, Jean. "L'Habitation rurale provençale: Le Vent et le soleil; Quelques remarques préliminaires." *Revue de géographie alpine* 31, no. 4 (1943): 525–33.

Merriman, John. *History on the Margins: People and Places in the Emergence of Modern France*. Lincoln: University of Nebraska Press, 2018.

Merry, Anne, and Bruno Lambert. *Flore des Calanques*. Les Sablonnières: Éditions du Fournel, 2014.

Méry, Louis-E. *Le Choléra à Marseille: Seconde invasion, 1835*. Marseille: Feissat et Demonchy, 1836.

Michelet, Jules. *History of France from the Earliest Period to the Present Time*. Translated by G. H. Smith. 19 vols. New York, 1851.

Michelet, Jules. *Oeuvres de M. Michelet*. Vol. 3. Brussels: Meline, Cans, 1840.

Michelot, Henri. *Portulan de la mer méditerranée ou guide des pilotes côtiers*. Marseille: Jean Mossy, 1805.

Ministère des Forêts. *Reboisement des montagnes (loi du 29 juillet 1860): Compte rendu des travaux de 1862*. Paris: Imprimerie Impériale, 1863.

Mistral, Frédéric. *Mes origines: Mémoires et récits*. Le Méjan: Actes Sud, 2008.

Mistral, Frédéric. *Mirèio: Pouèmo prouvençau / Mireille: Poème provençal*. Paris: Charpentier Libraire-Éditeur, 1868.

Mistral, Frédéric. *Lou Tresor dóu Felibrige ou Dictionnaire provençal-français*. 2 vols. Aix-en-Provence: Imprimerie Veuve Remondet-Aubin, 1878.

Mitchell, W. J. T. *Landscape and Power*. Chicago: University of Chicago Press, 2002.

Moore, Peter. *The Weather Experiment: The Pioneers Who Sought to See the Future*. New York: Farrar, Straus and Giroux, 2016.

Mondon, Bernard, and Steffen Lipp. *Petite anthologie du mistral*. Saint-Rémy-de-Provence: Éditions Equinoxe, 2004.

Monmonier, Mark. *Air Apparent: How Meteorologists Learned to Map, Predict, and Dramatize Weather*. Chicago: University of Chicago Press, 1999.

Murat, Pierre. "La Joliette, essor et sort pictural." *Marseille*, no. 244 (2014): 79–84.

Musard, Olivier, Laurence Le Dû-Blayo, Patrice Francour, Jean-Pierre Beurier, Eric Feunteun, and Luc Talassinos, eds. *Underwater Seascapes: From Geological to Ecological Perspectives*. London: Springer, 2014.

Nash, Linda. "The Agency of Nature or the Nature of Agency." *Environmental History* 10, no. 1 (January 2005): 67–69.

Nash, Linda. *Inescapable Ecologies: A History of Environment, Disease, and Knowledge*. Berkeley: University of California Press, 2006.

Nova, Alessandro. *The Book of the Wind: The Representation of the Invisible*. Montreal: McGill-Queen's University Press, 2011.

Novak, Barbara. *Nature and Culture: American Landscape and Painting, 1825–1875*. Oxford: Oxford University Press, 2007.

Obermann, Anika, Sophie Bastin, Sophie Belamari, Dario Conte, Miguel Angel Gaertner, Laurent Li, and Bodo Ahrens. "Mistral and Tramontane Wind Speed and Wind Direction Patterns in Regional Climate Simulations." *Climate Dynamics* 51, no. 3 (August 2018): 1059–76.

Oliver, John E. ed. *Encyclopedia of World Climatology*. New York: Springer, 2008.

Osborne, Michael A. *The Emergence of Tropical Medicine in France*. Chicago: University of Chicago Press, 2014.

Osborne, Michael A. "The Geographic Imperative in Nineteenth-Century French Medicine." In *Medical Geography in Historical Perspective*, edited by Nicolaas A. Rupke, 31–50. London: The Wellcome Trust for the History of Medicine at UCL, 2000.

Oxford English Dictionary Online. S.v. "mistral (*n.*)." Revised 2002. https://www.oed.com/dictionary/mistral_n.

Oxford English Dictionary Online. S.v. "sirocco (*n.*)." Revised 2002. https://www.oed.com/dictionary/sirocco_n.

Ozouf-Marignier, Marie-Vic. *La Formation des départements: La Représentation du territoire français à la fin du 18ᵉ siècle*. Paris: Éditions de l'École des hautes écoles en sciences sociales, 1989.

Pamard, Alfred. *L'Observatoire du Mont-Ventoux: Communication faite à l'Académie de Vaucluse, le 7 Février 1918*. Avignon: François Séguin, 1918.

Pansier, Paul. "Les Ascensions du Ventoux et la Chapelle de la Ste-Croix du XIVᵉ au XIXᵉ siècle." *Annales d'Avignon et du Comtat Venaissin* 18 (1932): 137–54.

Payn-Echalier, Patricia and Philippe Rigaud. *Pierre Giot: Un capitaine marin arlésien 'dans la tourmente.' Journal, livre de bord, correspondance, 1792–1816*. Aix-en-Provence: Presses universitaires de Provence, 2016.

Peck, Gunther. "The Nature of Labor: Fault Lines and Common Ground in Environmental and Labor History." *Environmental History* 11, no. 2 (April 2006): 212–38.

Perez, Louis A. *Winds of Change: Hurricanes and the Transformation of Nineteenth-Century Cuba*. Chapel Hill: University of North Carolina Press, 2001.

Pezet, Maurice. *La Provence sous le Mistral*. Arles-Raphèle: C. P. M. Culture Provençale et Méridionale, 1983.

Picon, Bernard. *L'Espace et le temps en Camargue: Histoire d'un delta face aux enjeux climatiques*. Arles: Actes Sud, 2020.

Pincetl, Stephanie. "Some Origins of French Environmentalism: An Exploration," *Forest and Conservation History* 37 (1993): 80–89.

Plumwood, Val. "The Concept of a Cultural Landscape: Nature, Culture and Agency in the Land." *Ethics and the Environment* 11, no. 2 (Fall-Winter 2006): 115–50.

Pöppel, Ernst, Mihai Avram, Yan Bao, Verena Graupmann, Evgeny Gutyrchik, Aline Lutz, Mona Park, et al. "Sensory Processing of Art as a Unique Window into Cognitive Mechanisms: Evidence from Behavioral Experiments and fMRI Studies." *Procedia—Social and Behavioral Sciences* 86 (October 2013): 10–17.

Pritchard, Sara. *Confluence: The Nature of Technology and the Remaking of the Rhône*. Cambridge, MA: Harvard University Press, 2011.

Pryor, John H. *Geography, Technology, and War: Studies in the Maritime History of the Mediterranean, 649–1571*. Cambridge: Cambridge University Press, 1992.

Pyne, Stephen. *Fire: Nature and Culture*. London: Reaktion Books, 2012.

Pyne, Stephen. *Vestal Fire: An Environmental History, Told through Fire, of Europe's Encounter with the World*. Seattle: University of Washington Press, 1997.

Quenet, Grégory. *Qu'est-ce que l'histoire environnementale?* Seyssel: Camp Vallon, 2014.

Rauch, F. A. *Harmonie Hydro-végétale et météorologique*. Paris: Frères Levrault, 1802.

Raymond, François. "Mémoire sur la topographie médicale de Marseille et son territoire; et sur celle des lieux voisins de cette ville." *Histoire de la Société royale de médecine*. Paris: Imprimerie de Philippe-Denys Pierres, 1780.

Reclus, Élisée. *The Ocean, Atmosphere, and Life*. Translated by B. B. Woodward. New York: Harper and Bros., 1873.

Redniss, Lauren. *Thunder and Lightning: Weather Past, Present and Future*. New York: Random House, 2015.

Reidy, Michael S. "Mountaineering, Masculinity, and the Male Body in Mid-Victorian Britain." *Osiris* 30, no. 1 (2015): 158–81.

Renan, Ernest. *"Qu'est-ce qu'une nation?": Conférence faite en Sorbonne, le 11 mars 1882*. Paris: Calmann Lévy, 1882.

Reynaud, Félix. *Ex-voto marins de Notre-Dame de la Garde*. Marseille: La Thune, 1996.

Robbins, Paul. *Political Ecology: A Critical Introduction*. Hoboken, NJ: Wiley-Blackwell, 2020.

Robinson, Peter J. "Ice and Snow in Paintings of Little Ice Age Winters," *Weather* 60, no. 2 (February 2005): 37–41.

Roucaute, Emeline, George Pichard, Eric Faure, and Manuela Royer-Carenzi. "Analysis of Causes of Spawning of Large-Scale, Severe Malarial Epidemics and Their Rapid Total Extinction in Western Provence, Historically a High Endemic Region of France (1745–1850)." *Malaria Journal* 13, no. 72 (2014): 1–42.

Roux, Marius. *Rapport général des travaux des conseils d'hygiène et de salubrité des trois arrondissements. Département des Bouches-du-Rhône*. Marseille: Typographie Barlatier-Feissat et Demonchy, 1851.

Ruffault, Julien, Vincent Moron, Ricardo M. Trigo, and Thomas Curt. "Daily Synoptic Conditions Associated with Large Fire Occurrence in Mediterranean France: Evidence for a Wind-Driven Fire Regime." *International Journal of Climatology* 37, no. 1 (2017): 524–33.

Schivelbusch, Wolfgang. *The Railway Journey: The Industrialization of Time and Space in the Nineteenth Century*. Berkeley: University of California Press, 2014.

Schulten, Susan. *Mapping the Nation: History and Cartography in Nineteenth-Century America*. Chicago: University of Chicago Press, 2012.

Scott, James C. *Seeing Like a State: How Certain Schemes to Improve the Human Condition Have Failed*. New Haven, CT: Yale University Press, 1998.

Serre, Rodolphe. *Port de Marseille: Ses eaux renouvelées en 36 heures*. Marseille: Petit Provençal, 1884.

Sewell, William H., Jr. *Structure and Mobility: The Men and Women of Marseille, 1820–1870*. Cambridge: Cambridge University Press, 2009.

Shapin, Steven. *Never Pure: Historical Studies of Science as if It Was Produced by People with Bodies, Situated in Time, Space, Culture, and Society, and Struggling for Credibility and Authority*. Baltimore: Johns Hopkins University Press, 2010.

Silverman, Debora. *Van Gogh and Gauguin: The Search for Sacred Art*. New York: Farrar, Straus and Giroux, 2000.

Singer, Brian C. J. "Cultural versus Contractual Nations: Rethinking Their Opposition." *History and Theory* 35, no. 3 (October 1996): 309–37.

Smail, Daniel Lord. *On Deep History and the Brain*. Berkeley: University of California Press, 2008.

Smets, Bas, and Eliane Leroux. "Luma Arles: The Making of a Climate." *TLmag36 Extended: All Is Landscape*, January 21, 2022. https://tlmagazine.com/luma-arles-the-making-of-a-climate/.

Smith, Philip Chadwick Foster. *The Artful Roux: Marine Painters of Marseille*. Salem, MA: Peabody Museum of Salem, 1978.

Société royale de médecine. *Histoire de la Société royale de médecine*. Paris: Imprimerie de Philippe-Denys Pierres, 1780.

Solana, Guillermo, ed. *Gauguin and the Origins of Symbolism*. Madrid: Philip Wilson Publishers, 2004.

Solomon, Deborah. "Van Gogh and the Consolation of Trees." *New York Times*, May 14, 2023. https://www.nytimes.com/2023/05/11/arts/design/van-gogh-cypresses-met-museum.html.

Soubiran, Jean-Roger. *Le Paysage provençal et l'école de Marseille avant l'impressionnisme, 1845–1874*. Toulon: Musée de Toulon, 1992.

Soubiran, Jean-Roger. *Paysages provençaux de Loubon à Ambrogiani*. Marseille: Éditions Jeanne Lafitte, 2013.

Stein, Susan Alyson. *Van Gogh's Cypresses*. New York: Metropolitan Museum of Art, 2023.

Stendhal. *Mémoires d'un touriste*. Vol. l. Paris: Michel Lévy Frères, 1854.

Strabo. *Géographie de Strabon: Traduite de grec en français*. Vol. 2. Paris: Imprimerie Impériale, 1809.

Strauss, Sarah. "An Ill Wind: The Foehn in Leukerbad and Beyond," *Journal of the Royal Anthropological Institute* 13 (2007): 165–81.

Streever, Bill. *And Soon I Heard a Roaring Wind: A Natural History of Moving Air*. New York: Little, Brown, 2016.

Surkis, Judith. *Sexing the Citizen: Morality and Masculinity in France, 1870–1920*. Ithaca, NY: Cornell University Press, 2006.

Tabeaud, Martine. "Climats urbains: Savoirs experts et pratiques sociales." *Ethnologie française* 40, no. 4 (October-December 2010): 685–94.

Tabeaud, Martine, Benjamin Lysaniuk, Nicolas Schoenenwald, and Jérôme Buridant. "Le Risque 'coup de vent' en France depuis le XVIe siècle." *Annales de Géographie* 118, no. 667 (2009): 318–31.

Takeda, Junko Thérèse. *Between Crown and Commerce: Marseille and the Early Modern Mediterranean*. Baltimore: Johns Hopkins University Press, 2011.

Thiesse, Anne-Marie. *Ils apprenaient la France: L'Exaltation des régions dans le discours patriotique*. Paris: Éditions de la Maison des Sciences de l'Homme, 1997.

Thomas, Greg M. *Art and Ecology in Nineteenth-Century France: The Landscapes of Théodore Rousseau*. Princeton, NJ: Princeton University Press, 2000.

Thomson, Belinda, ed. *Gauguin: Maker of Myth*. London: Tate Publishing, 2010.

Tissander, Gaston. "L'Observatoire météorologique du Mont Ventoux." *La Nature*, no. 599. November 22, 1884.

Togunov, Ron, Andrew E. Derocher, and Nicholas J. Lunn. "Windscapes and Olfactory Foraging in a Large Carnivore." *Scientific Reports* 7, no. 46332 (2017). https://doi.org/10.1038/srep46332.

Turnbull, David. "Reframing Science and Other Local Knowledge Traditions." *Futures* 29 (1997): 551–62.

Valencius, Conevery Bolton. *The Health of the Country: How American Settlers Understood Themselves and Their Land*. New York: Basic Books, 2002.

van Gogh, Vincent. *Lettres à son frère Théo*. Translated by Louis Roëdlant. Paris: Gallimard, 1988.

Vanhoenacker, Mark. "The Wind Cries . . . Oe?" *New York Times*, December 13, 2013. https://www.nytimes.com/2013/12/24/science/earth/the-wind-cries-oe.html.

Vaulabelle, Alfred de. "Un nouvel observatoire: L'Observatoire du Mont Ventoux." *Magasin pittoresque*, no. 10 (1886): 164–66.

Veale, Lucy, Georgina Endfield, and Simon Naylor. "Knowing Weather in Place: The Helm Wind of Cross Fell." *Journal of Historical Geography* 45 (2014): 25–37.

Vedder, Aline, Lukasz Smigielski, Evgeny Gutyrchik, Yan Bao, Janusch Blautzik, Ernst Pöppel, Yuliya Zaytseva, and Edmund Russell. "Neurofunctional Correlates of Environmental Cognition: An fMRI Study with Images from Episodic Memory." *PLoS One* 10, no. 4 (April 2015): 1–11.

Vellekoop, Marije. *Van Gogh at Work*. Brussels: Mercatorfonds, 2013.

Vence, Jules. *Construction et manoeuvre des bateaux et embarcations à voilure latine*. Paris: Augustin Challamel, 1897.

Vetter, Jeremy. *Field Life: Science and the American West during the Railroad Era*. Pittsburgh: University of Pittsburgh Press, 2016.

Vidal de la Blache, Paul. *Géographie universelle*. 7 vols. Paris: Armand Colin, 1934.

Vidal de la Blache, Paul. *Principles of Human Geography*. Translated by Milicent Todd Bingham. New York: Henry Holt, 1926.

Viegnes, Michel, ed. *Imaginaires du vent*. Paris: Imago, 2003.

Watson, Lyall. *Heaven's Breath: A Natural History of the Wind*. New York: New York Review Books, 2019.

Weidensaul, Scott. *Living on the Wind: Across the Hemisphere with Migratory Birds*. New York: North Point Press, 1999.

White, Richard. "'Are You an Environmentalist or Do You Work for a Living?': Work and Nature." In *Uncommon Ground: Rethinking the Human Place in Nature*, edited by William Cronon, 171–85. New York: W. W. Norton, 1996.

White, Sam. *A Cold Welcome: The Little Ice Age and Europe's Encounter with North America*. Cambridge, MA: Harvard University Press, 2017.

Williams, James C. "Sailing as Play." *Icon* 19 (2013): 132–92.

Worster, Donald. *Nature's Economy: A History of Ecological Ideas*. Cambridge: Cambridge University Press, 1994.

Wrigley, E. A. *The Path to Sustained Growth: England's Transition from an Organic Economy to an Industrial Revolution*. Cambridge: Cambridge University Press, 2016.

Wylie, Laurence. *Village in the Vaucluse*. Cambridge, MA: Harvard University Press, 1974.

Zaretsky, Robert. *Cock and Bull Stories: Folco de Baroncelli and the Invention of the Camargue*. Lincoln: University of Nebraska Press, 2004.

Zeng, Zhenzhong, Alan D. Ziegler, Timothy Searchinger, Long Yang, Anping Chen, Kunlu Ju, Shilong Piao, et al. "A Reversal in Global Terrestrial Stilling and Its Implications for Wind Energy Production." *Nature Climate Change* 9 (December 2019): 979–85.

Zola, Émile. *Doctor Pascale*. Translated by Julie Rose. Oxford: Oxford University Press, 2020.

Zola, Émile. *Naïs Micoulin*. Paris: G. Charpentier, 1883.

INDEX